Lecture Notes in Electrical Engineering

Volume 353

About this Series

"Lecture Notes in Electrical Engineering (LNEE)" is a book series which reports the latest research and developments in Electrical Engineering, namely:

- Communication, Networks, and Information Theory
- Computer Engineering
- Signal, Image, Speech and Information Processing
- Circuits and Systems
- Bioengineering

LNEE publishes authored monographs and contributed volumes which present cutting edge research information as well as new perspectives on classical fields, while maintaining Springer's high standards of academic excellence. Also considered for publication are lecture materials, proceedings, and other related materials of exceptionally high quality and interest. The subject matter should be original and timely, reporting the latest research and developments in all areas of electrical engineering.

The audience for the books in LNEE consists of advanced level students, researchers, and industry professionals working at the forefront of their fields. Much like Springer's other Lecture Notes series, LNEE will be distributed through Springer's print and electronic publishing channels.

More information about this series at http://www.springer.com/series/7818

Yue Zhang

An Introduction to Python and Computer Programming

Springer

Yue Zhang
Singapore University of Technology
and Design
Singapore
Singapore

ISSN 1876-1100 ISSN 1876-1119 (electronic)
Lecture Notes in Electrical Engineering
ISBN 978-981-10-1243-3 ISBN 978-981-287-609-6 (eBook)
DOI 10.1007/978-981-287-609-6

Springer Singapore Heidelberg New York Dordrecht London

Softcover reprint of the hardcover 1st edition 2015

Printed on acid-free paper

Springer Science+Business Media Singapore Pte Ltd. is part of Springer Science+Business Media (www.springer.com)

Preface

I have been teaching Python to first-year engineering students in Singapore University of Technology and Design. As a powerful and high level programming language, Python is ideal for the introduction to programming, which is useful for all engineering disciplines. In my teaching, I constantly find the need for ways to make my explanations as simple as possible. A concept, no matter how simple it seems to us computer scientists, may raise profound questions among students. For example, why is $x = x + 1$ a valid statement but not equivalent to the contradicting equation $0 = 1$? How can three lines of static code lead to a thousand statements being executed dynamically?

A student who asks such questions can be a very good student. Once the basic concepts are explained, she can manage programming effectively. On the other hand, few books on Python touch on the simple yet fundamental discussions that have been accumulated into my lecture notes. This has become my motivation for writing this book, a very gentle introduction to Python and programming.

From teaching experiences in University of Oxford, University of Cambridge, and Singapore University of Technology and Design, I find three components the most important for teaching programming. The first is syntax, the basic knowledge of how to write a correct program. As with most books on the Python language, it is an important component of this book. The second is the underlying mechanism of dynamic execution, and the underlying representation of data types. This is an aspect of programming that can easily be overlooked, which leads to common mistakes in programming. I make abstractions to the most important parts of the Python kernel, such as the binding table, object structures, and statement execution, showing them in figures, so that students can be confident of what happens underneath when a program is executed.

The third component is problem solving, which is the target of program design. The goal of learning programming is to solve problems, and it is important to associate programming concepts to problem solving when they are introduced. There are often various typical ways to approach common problems, and I try to introduce them along with the introduction to the programming language itself.

In addition, I include the most basic concepts in computer science into the introduction to Python, including information theory, computer architecture, data structure and algorithms, numerical analysis, and program design thinking such as functional programming and object-oriented programming. All these concepts are necessary backgrounds on which Python operates, and through which students can acquire a basic idea of computer science. I try to merge them into relevant sections, and make my introduction most simple and gentle.

In terms of Python syntax and libraries, I prioritize the former over the latter, introducing the full Python syntax and the most important libraries. The goal is to introduce Python to students with no programming or computer science background, equipping them with knowledge of the programming language and its underlying mechanisms, so that they can learn the usage of additional libraries with little difficulty, by consulting the Python documentation.

It took me over a year to transform my lecture notes into the book, during which many people gave me valuable help. Special thanks to Qiujing Xu, Likun Qiu, Chen Lü, Yanan Lu, Chingyun Chang, Yijia Liu, Zhongye Jia, Bo Chen, Jie Yang, Meishan Zhang, Xiao Ding, Hao Zhou, and Ji Ma for helping me to draw the figures, and enter handwritten text in some chapters, to Haoliang Qi for inviting me to give a guest lecture series and encouragement to write up my lecture notes, and to all the students who gave me valuable feedback.

Yue Zhang

Contents

Chapter 1
An Introduction to Python and Computer Programming

1.1 Introduction

Computers play a very important role in the digital world, and in our daily lives. They are central to smart TVs, phones, homes cars, metros, airplanes, robots, factories and many other digital systems. We use computers through these systems every day.

At work, computers have become an indispensible tool to more and more people. Fund managers use computers for analytical trading; architects use computers to make and verify building design; linguists use computers to study and annotate text corpora; natural scientists use computers to simulate large experiments in physics and chemistry; supermarkets use computers to plan and schedule the most efficient home delivery times and routes. As a result, fundamental knowledge on effectively utilizing computers is useful for all industries and disciplines.

Computers integrate **hardware** and **software**. Hardware refers to physical components, such as chips, keyboards and monitors, while software refers to operating systems and application programs, such as *Office*, *Firefox* and *iTunes*. Software operates on hardware to provide useful functionalities to users, such as text editing, Internet browsing and gaming. On the other hand, hardware provides fundamental support to software, by offering a set of *basic instructions*, which software combines to achieve complex functionality.

Programming is the design of software using a set of basic instructions. **Programming languages** abstract hardware instructions into basic **statements**, which are much easier to understand and handle. Using a programming language, one can write complex software, achieving tailored functionalities by combination of basic statements. This is the ultimate way of utilizing a computer, and is necessary when existing software fails to satisfy the requirement, such as the need to compute novel equations, or simulate specific experiments in an engineering discipline.

Y. Zhang, *An Introduction to Python and Computer Programming*,
Lecture Notes in Electrical Engineering 353, DOI 10.1007/978-981-287-609-6_1

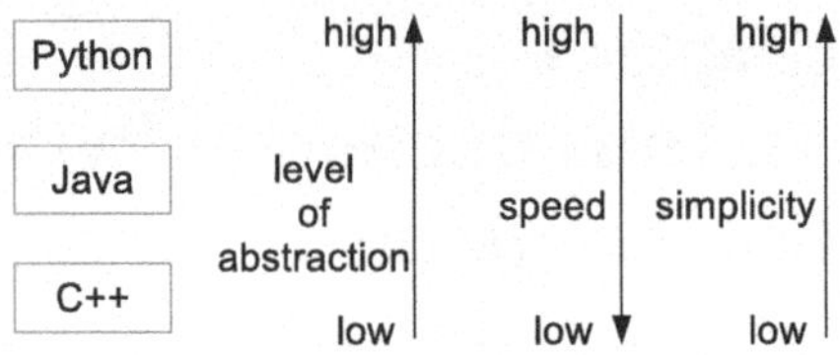

Fig. 1.1 Comparison of Python, Java and C++

1.1.1 Python and Computer Programming

Python was invented in the 1990s, named after the comedy show *Monte Python's Flying Circus*. Due to its simplicity and elegance, it gained popularity among programmers quickly, and was adopted by large companies, such as Google Inc. Compared with other programming languages, such as Java and C++, Python is particularly useful for learning computer programming.

Figure 1.1 shows a comparison between Python, Java and C++, three popular programming languages. Among the three, C++ was invented the earliest. Its statements were the closest to hardware instructions. As a result, C++ programs run very fast, and can be optimized in many ways. However, mastering C++ requires knowledge in computer hardware. Java was invented decades after C++. It offers higher levels of abstraction compared to C++, making it easier to program. As a trade-off, Java programs run slower than C++. Python takes one step further in abstracting away hardware details, offering a conceptually very simple system for statement execution. As a result, Python is the easiest to learn and fully master, but Python programs are typically the slowest to run.

In terms of functionalities, Python is as powerful as Java and C++. It supports all the functionalities that typical computer hardware supports. In addition, libraries provided by Python and open-source contributers enrich the power of Python, making it the most convenient choice for many applications, such as text processing. Python libraries such as *numpy* and *scipy* are implemented in C++ in the lower level, offering very fast ways to perform scientific computation using Python.

Problem-solving skills are as important as knowledge of programming languages to computer programming. It is therefore important to learn common approaches to typical problems when learning computer programming. Python supports virtually all the programming paradigms that Java and C++ supports, and even more, making it convenient to apply flexible problem-solving techniques using the Python programming language.

1.2 Preliminaries

As preliminary knowledge to Python programming, it is helpful to know the basic structure of a computer, the operating system, the file system, and how to install and configure Python. This section introduces these backgrounds, so that it is easier to embark on Python programming from the next chapter.

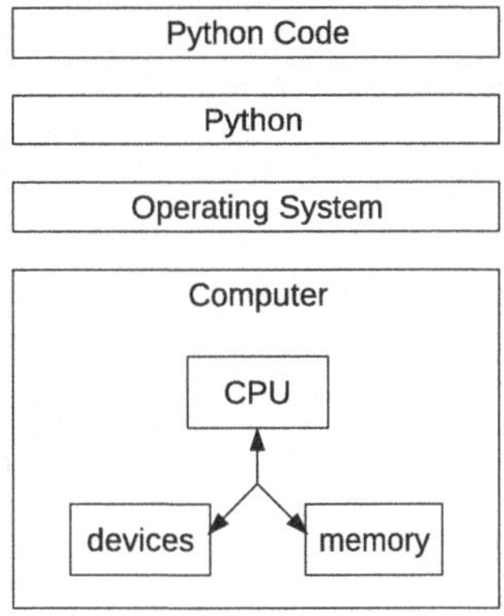

Fig. 1.2 The structure of a computer

1.2.1 The Computer

The structure of a computer and the mechanism in which Python code is executed are illustrated in Fig. 1.2. On the bottom of the figure is the hardware computer, which is physically tangible. It consists of three main components: the **central processing unit** (CPU), the **memory** and **devices**. They will be discussed in the next chapter. On top of the computer runs an **operating system**, such as *Microsoft Windows*, *Linux* and *Mac OS*. Operating systems connect computer hardware and software, providing basic interfaces to software programs for controlling the hardware. For example, the *file system* is an interface that operating systems provide for organizing data in the hard drive (hardware). Browsers, text editors, music players, and the vast majority of other application programs are managed by the operating system.

As shown in Fig. 1.2, Python is an application program running on top of the operating system, just like a browser or a text editor. Its main function, however, is to execute generic Python code, rather than performing a specific task, which browsers and text editors do. IDLE is another application program, which provides an interactive interface for Python. On the top of the figure is Python code, which is executed by the Python application. Given a piece of Python code, or a **Python program**, Python executes it by **interpreting** the code into computer instructions, executed on computer hardware via the operating system. In the following chapters, how various types of Python statements are interpreted will be discussed in detail.

1.2.2 The File System

Operating systems organize data in a hard drive as **files**. Each file contains a certain type of data. For example, a music track can be stored as an *mp3* file, a video clip can be stored as an *mpeg* file, and a text document can be stored as a *txt* file. Application programs are also files. In *Windows*, for example, application programs such as

Firefox are stored as executable (*exe*) files. Python programs are a special type of text document, stored as *py* files.

The type of a file, such as as *mp3* and *txt*, is typically reflected by the last part of the file name, which is called the **extension name**. The extension name is separated from the main file name by a dot (.). For example, in the file name "*readme.txt*", the main file name is "*readme*" and the extension name is "*txt*", indicating that the corresponding file is a text file. In the file name "*hello.py*", the main file name is "*hello*", and the extension name is "*py*", indicating that the file is a Python source file.

Files are organized using **folders**, which are containers of files and other folders. In a file system, folders are organized in a hierarchy. There is only one folder that is not contained by another folder, and it is called the **root** folder. Except for the root folder, each folder belongs to another folder, which is typically called the **parent folder** of the folder. Such a hierarchical structure is called a **tree**. An example is shown in Fig. 1.3, where the folders and files form an upside-down tree from the root folder.

In the tree structure of a file system, each file can be identified by the **path** from the root folder to the file. For example, in Fig. 1.3, *file1* can be identified by the path *root* → *folder1* → *file1*, and *file3* can be identified by the path *root* → *folder1* → *folder4* → *folder7* → *file3*. The use of path names to identify files avoids name clashes between different folders. For example, even if *file6* is renamed as *file1* in Fig. 1.3, it is still different from *root* → *folder1* → *file1*.

In *Windows*, back slashes (\) are used as the path deliminator symbol. In *Linux* and *Max OS*, slashes (/) are used as the path deliminator. For example, the path for *file6* in Fig. 1.3 is written as *folder3**folder6**file6* in *Windows*, and */folder3/folder6/file6* in *Linux*. In all these cases the root folder is written implicitly as the start of the path.

In *Windows*, a hard drive can be split into several **volumes**, each having its own tree structure of folders. *Windows* volumes are denoted by letters, starting from '*C*', followed by a colon (:). For example, '*C:\folder\file1*' represents a file in the volume '*C:*', and is different from the file '*D:\folder1\file1*'.

Fig. 1.3 Folder hierarchy

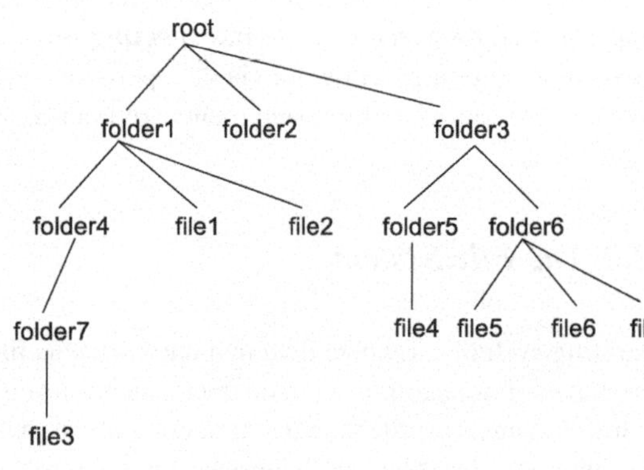

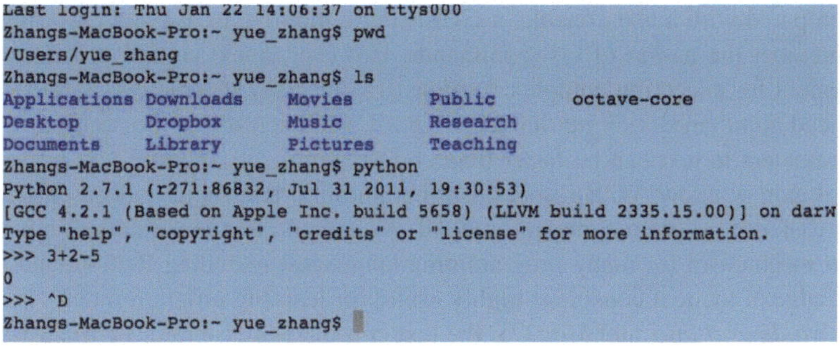

Fig. 1.4 Example text console *Terminal* on a *Mac OS*

1.2.3 Text User Interfaces to Operating Systems

An operating system consists of many components, and one way to categorize operating system components is to put them into two types. The first type of OS components operates the computer hardware; it is called the **kernel** in *Linux* systems, for performing the "core" functionalities. The second type of OS components provides a user interface; it is called the **shell** in *Linux* systems.

OS user interfaces have evolved from text user interface to **graphical user interface** (GUI). A text user interface is also called a **text console** (Fig. 1.4). It works by repeatedly displaying a **prompt message** to the user and waits for a line of user command. After receiving a command, it executes the command, displays results and feedbacks in text format, and prompts for a next command. On the other hand, a GUI (Fig. 1.5) displays information in graphical windows and message boxes, and receives user command via buttons, menu items and dialog boxes.

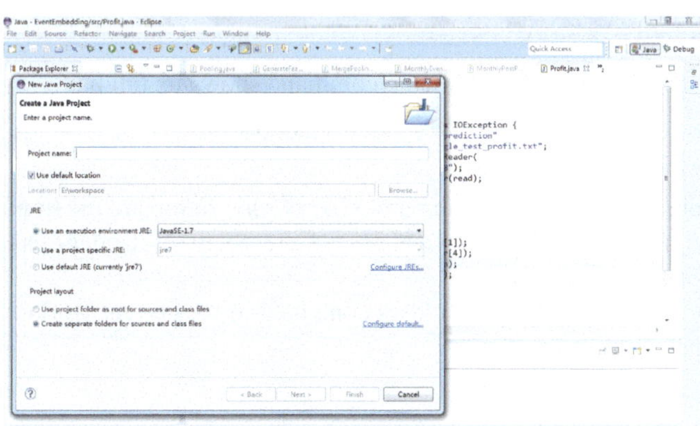

Fig. 1.5 Example graphical user interface

Compared with a text console, a GUI is more intuitive to use, not requiring one to remember the names of text commands. However, a text console is often more convenient for executing complex commands with many parameters, because such advanced commands may not have GUI at all, and even if they do, entering a row of parameters in text can be faster than choosing and entering into text boxes in a large dialog window. With respect to programming, text input and output can be much easier to program compared to GUI. As a result, text is the default input and output mechanism for many programming languages, including Python, and basic knowledge of the text console is highly useful for learning programming.

In *Windows*, *Linux* and *Mac OS*, the text console is a program. In *Windows*, the program is named '*Command Line*'; in *Linux*, the program is named '*Terminal*' or '*Console*'; in *Mac OS*, the program is named '*Terminal*' (Fig. 1.4). Because the text console commands of *Linux* and *Mac OS* are mostly similar, they are not introduced separately in this book.

After being launched, *Command Line* can show the following message:

```
C:\Users\yue_zhang>
```

After being launched, *Terminal* can show the following message:

```
Zhangs-MacBook-Pro:~ yue_zhang$
```

In the examples above, 'C:*Users**yue_zhang*>' and '*Zhangs-MacBook-Pro:* ~ yue_zhang$' are the prompt messages, after which a command can be entered. Each text command finishes with a new line (the *Enter* key).

In a text console, all commands are entered in a current **working folder**, which is a folder in the file system. By default, all the commands that are entered into the text console operates on the current working folder. Without a full path being specified, all file names entered into the text console also refer to files in the current working folder by default. The current working folder is typically reflected by the prompt message. In *Windows*, the following command shows its full path:

```
C:\Users\yue_zhang> cd
C:\Users\yue_zhang
```

In *Linux* and *Mac OS*, the following command shows its full path:

```
Zhangs-MacBook-Pro:~ yue_zhang$ pwd
/Users/yue_zhang
```

By default, C:*Users**yue_zhang* and */Users/yue_zhang* are the **home folder** of the user *yue_zhang* under *Windows* and *Mac OS*, respectively. Under *Linux*, the home folder can be */home/yue_zhang*.

One can list out the files and folders contained in the current working folder. In *Windows*, the command is *dir*:

```
C:\Users\yue_zhang>dir
2015/03/30   15:30    <DIR>         .
2015/03/30   15:30    <DIR>         ..
2015/03/12   10:15    <DIR>         Contacts
2015/03/30   14:49    <DIR>         Desktop
2015/03/12   10:15    <DIR>         Documents
```

```
2015/03/12  10:15    <DIR>          Downloads
2015/03/12  10:15    <DIR>          Favorites
```

In *Linux* and *Mac OS*, the command is *ls*:

```
Zhangs-MacBook-Pro:~ yue_zhang$ ls
Applications     Downloads    Movies       Public
   octave-core
Desktop          Dropbox      Music        Research
Documents        Library      Pictures     Teaching
```

All the commands are by default executed in the current working folder. For example, the command '*mkdir <folder_name>*' creates a new folder named '*<folder_name>*' in the current folder. An example in *Windows* is shown below.

```
C:\Users\yue_zhang>mkdir newfolder

C:\Users\yue_zhang>dir
2015/03/30  15:30    <DIR>          .
2015/03/30  15:30    <DIR>          ..
2015/03/12  10:15    <DIR>          Contacts
2015/03/30  14:49    <DIR>          Desktop
2015/03/12  10:15    <DIR>          Documents
2015/03/12  10:15    <DIR>          Downloads
2015/03/12  10:15    <DIR>          Favorites
2015/03/30  15:26    <DIR>          newfolder
```

The command '*rmdir <folder_name>*' removes the empty folder '*<folder_name>*' from the current folder:

```
C:\Users\yue_zhang>rmdir newfolder

C:\Users\yue_zhang>dir
2015/03/30  15:30    <DIR>          .
2015/03/30  15:30    <DIR>          ..
2015/03/12  10:15    <DIR>          Contacts
2015/03/30  14:49    <DIR>          Desktop
2015/03/12  10:15    <DIR>          Documents
2015/03/12  10:15    <DIR>          Downloads
2015/03/12  10:15    <DIR>          Favorites
```

The commands *mkdir* and *rmdir* apply to *Linux* and *Mac OS* also.

One can change the current working folder by using the command '*cd* <folder_name>', which enters the folder *<folder_name>*. For example,

```
C:\Users\yue_zhang>dir
2015/03/30  15:30    <DIR>          .
2015/03/30  15:30    <DIR>          ..
2015/03/12  10:15    <DIR>          Contacts
2015/03/30  14:49    <DIR>          Desktop
2015/03/12  10:15    <DIR>          Documents
2015/03/12  10:15    <DIR>          Downloads
2015/03/12  10:15    <DIR>          Favorites
C:\Users\yue_zhang>cd Contacts

C:\Users\yue_zhang\Contacts>
```

Two **special folder names** are '.' and '..', which represent the current working folder and its parent folder, respectively. 'cd .' does nothing, for the working folder is changed to the same folder. 'cd ..' enters the parent folder, the folder one level up towards the root.

```
C:\Users\yue_zhang>cd
C:\Users\yue_zhang

C:\Users\yue_zhang>cd ..

C:\Users>
```

In all the examples above, *<folder_name>* is a folder name in the current working folder. This type of path is called a **relative path**. On the other hand, **absolute paths**, which specify a folder from the root, can also be used for *<folder_name>*. For example,

```
C:\Users\yue_zhang>cd
C:\Users\yue_zhang

C:\Users\yue_zhang>cd \Users\yue_zhang\Desktop

C:\Users\yue_zhang\Desktop>
```

The syntax of the *cd <folder>* command is the same in *Linux* and *Mac OS* as in *Windows*. However, if *cd* is used without specifying *<folder>*, the functionality is showing the current working directory on *Windows*, while entering the home folder on *Linux* and *Mac OS*.

1.2.4 The Python Application Program

Python programs are executed by **the Python application program**, or **Python** in short, which is a software program like *TextEdit* and *Firefox*. Python can be downloaded from the official Python website www.python.org, and installed into a computer. There are two commonly used dialects of Python: Python 2 and Python 3. This book uses Python 2, in which most existing Python programs are written. However, the underlying mechanisms are mostly the same for both Python 2 and Python 3, and they are only mildly different in syntax for certain functionalities.

The Python installer program allows the specification of a path, in which Python is installed. By default, the folder is *C:\Python2.7* on *Windows*, where 2.7 is the version number of Python. After installation, the executable Python application is the file:

```
C:\Python2.7\python.exe
```

On *Linux* and *Mac OS*, Python is typically pre-installed, and can be found at

```
/usr/bin/python2.7
```

Python can be executed by setting the current working folder to the folder containing Python:

```
c:\Users\yue_zhang>cd \Python2.7
C:\Python2.7>python
Python 2.7.1 (r271:86832, Jul 31 2011, 19:30:53) [MSC
    v.1500 32 bit (Intel)] on win32
Type"help", "copyright", "credits" or "license" for
    more information.
>>>
```

By typing '*python*', *Windows* executes the file *python.exe* in the current working folder, which leads to the interactive execution mode of Python.

On *Linux* and *Mac OS*, Python can be executed by typing '*./Python2.7*' in the current working folder.

```
Zhangs-MacBook-Pro:~ yue_zhang$ cd /usr/bin
Zhangs-MacBook-Pro:~ yue_zhang$ ./python2.7
Python 2.7.1 (r271:86832, Jul 31 2011, 19:30:53)
[GCC 4.2.1 (Based on Apple Inc. build 5658) (LLVM
    build 2335.15.00)] on linux2
Type "help", "copyright", "credits" or "license" for
    more information.
>>>
```

Note that './' must be used when executing a program in the current working folder in *Linux* and *Mac OS*. To exit the interactive Python application, press and hold the *[Ctrl]* key and press the *D* key (i.e. press *Ctrl-D*).

```
(continued from above)
>>> ^D
Zhangs-MacBook-Pro:~ yue_zhang$
```

1.2.5 Python and Environment Variables

The content of this section is not typically necessary to know, but can be useful when customizing Python.

Environment variables are a set of settings that an operating system uses to configure itself. For *Windows*, the current environment variables can be listed and modified by clicking *Start → Control Panel → System → Advanced system settings*. In the *Advanced* ***tab***, select *Environment Variables*, as shown in Fig. 1.6.

For *Linux* and *Mac OS*, the current environment variables can be viewed and modified by editing the file *$PATH*.

The ***PATH*** **environment variable** stores a list of paths from which the operating system searches for an executable program automatically. If the specified executable is not in the current working folder (*Windows*), or the path (e.g. './'; '*/Desktop/*') is not specified (*Linux* and *Mac OS*), the operating system will search the paths in the *PATH* environment variable for the executable.

Python is typically added to the *PATH* environment variable when installed. Therefore, it should be possible to execute *python* at any path. In the rare cases when the operating system cannot find the python executable program, the path in which

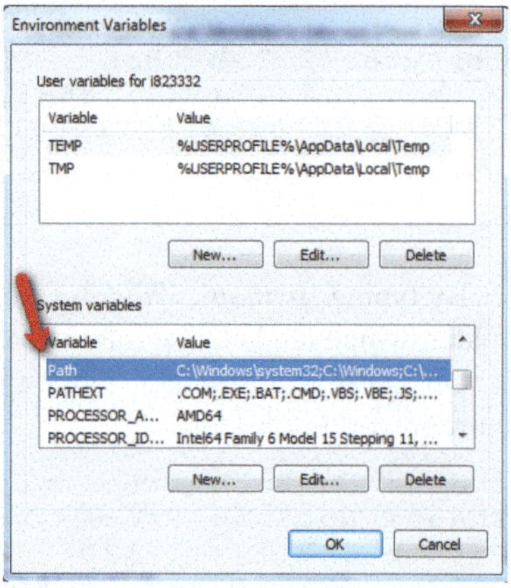

Fig. 1.6 Viewing and setting environment variables for *Windows*

Python is installed (e.g. *C:\Python 2.7*) can be added to the list of folders in the *PATH* environment variable. A restart of the operating system may be necessary for the change to take effect.

The ***PYTHON_PATH*** **environment variable** is a second environment variable relevant to Python. It also stores a list of paths that are useful to Python specifically, rather than for the operating system. In particular, when Python loads library modules (Chap. 3), it will search the paths in *PYTHON_PATH* for them.

On *Windows*, *PYTHON_PATH* can be set by following Fig. 1.6. On *Linux* and *Mac OS*, it can be set by modifying the file *.bash_profile* under the home folder, adding the line

```
export PYTHON_PATH=$PYTHON_PATH:path1:path2:path3
```

where *path1*, *path2* and *path3* are module paths.

Exercises

1. Which of the following can be achieved using Python programming?
   (a) Mathmatical calculations;
   (b) A *Windows* GUI application;
   (c) Controlling a robotic arm;
   (d) Web crawling;
   (e) A machine-leaning system that can fly a helicopter;
   (f) Simulation of large-scale experiments in physics;
   (g) A game in Android
   (h) A operating system from scratch.
2. What are the three main components of a computer?
3. List three programming languages and compare their differences. Why are there many programming languages?
4. How can text user interfaces be launched on *Windows*, *Linux* and *Mac OS*?

5. Perform the following in a text user interface.
   (a) Show the current working folder;
   (b) Enter the home folder;
   (c) Enter the *Desktop* folder;
   (d) Create a new folder under the name *New* under the *Desktop* folder, and verify that it appears on the graphical desktop;
6. Launch the *python* program, and then exit it.

Chapter 2
Using Python as a Calculator

One of the most important tasks that a computer performs is mathematical computation. In fact, the computation of mathematical functions had been the driving motivation for the invention of 'computing machines' by pioneer researchers. As other programming languages do, Python provides a direct interface to this fundamental functionality of modern computers. Naturally, an introduction of Python could start by showing how it can be used as a tool for simple mathematical calculations.

2.1 Using Python as a Calculator

The easiest way to perform mathematics calculation using Python is to use IDLE, the interactive development environment of Python, which can be used as a fancy calculator. To begin with the simplest mathematical functions, including integral addition, subtraction, multiplication and division, can be performed in IDLE using the following mathematical **expressions**[1]:

```
>>> 3+5          # addition
8
>>> 3-2          # subtraction
1
>>> 6*7          # multiplication
42
>>> 8/4          # division
2
```

As can be seen from the examples above, a simple Python expression is similar to a mathematical expression. It consists of some numbers, connected by a mathematical **operator**. In programming terminology, number constants (e.g. 3, 5, 2) are called **literals**. An operator (e.g. $+$, $-$, $*$) indicates the mathematical function between

[1]The content after the # symbols are comments and can be ignored when typing the examples. Details about comments are given in Chap. 3 **or** the rest of the book >>> indicates an IDLE command.

Y. Zhang, *An Introduction to Python and Computer Programming*,
Lecture Notes in Electrical Engineering 353, DOI 10.1007/978-981-287-609-6_2

its **operands**, and hence the **value** of the expression (e.g. $3 + 5$). The process of deriving the value of an expression is called the **evaluation** of the expression. When a mathematical expression is entered, IDLE automatically evaluates it and displays its value in the next line.

Since expressions themselves represent values, they can be used as operands in longer *composite expressions*.

```
>>> 3+2-5+1
1
```

In the example above, the expression 3+2 is evaluated first. Its value 5 is combined with the next literal 5 by the − operator, resulting in the value 0 of the composite expression $3 + 2 - 5$. This value is in turn combined with the last literal 1 by the + operator, ending up with the value 1 for the whole expression.

In the example, operators are applied from left to right, because + and − have the same priority. In an expression that contains more than one operators, not necessarily all operators have the same priority. For example,

```
>>> 3+2*5-4
9
```

The expression above evaluates to 9, because the multiplication ($*$) operator has a higher priority compared with the + and − operators. The order in which operators are applied is called **operator precedence**, and mathematical operators in Python follow the natural precedence by the mathematical function. For example, multiplication ($*$) and division (/) have higher priorities than addition (+) and subtraction (−). An operator that has higher priority than multiplication is the power operator ($**$).

```
>>> 5 ** 2            # power
25
```

In general, $a ** b$ denotes the value of a to the power of b.

Similar to mathematical equations, Python allows the use of brackets (i.e. '(' and ')') to manually specify the order of evaluation. For example,

```
>>> (3+2)*(5-4)      # brackets
5
```

The value of the expression above is 5 because the bracketed expressions $3 + 2$ and $5 - 4$ are evaluated first, before the $*$ operator is applied.

In the examples above, an operator connects two literal operands, and hence they are called *binary operators*. An operator can also be *unary*, taking only a single operand. An example unary operator is −, which takes a single operand and negates the number.

```
>>> -(5*1)           # negation
-5
```

There is also a *ternary operator* in Python, which takes three operands. It will be introduced in a later chapter.

Until this point, all the mathematical expressions have been integral, with the values of literal operands and expressions being integers. Take the division operator (/) for example,

```
>>> 5/2            # division with integer operands
2
```

The result of the integral division operation is the quotient 2, with the fractional part 1 discarded. To find the remainder of integer division, the *modulo operator* (%) can be used.

```
>>> 5%2            # modulo
1
```

2.1.1 Floating Point Expressions

So for all the expressions in this chapter are integer expressions, of which all the operands and the value are integers. However, for the expressions 5/2, sometimes the real number 2.5 is a more appropriate value. In computer science, real number are typically called **floating point number**. To perform floating point arithmetics, at least one floating point number must be put in the expression, which results in a **floating point expression**. For example,

```
>>> 5.0/2         # floating point division
2.5
>>> 5/2.0         # floating point division
2.5
>>> 25 ** 0.5     # floating point power
5.0
```

The last example above calculates the positive square root of 25. Regardless of operands, when all the numbers in a Python expression are integers, the expression is an integer expression, and the value of the expression itself is an integer. However, when there is at least one floating point number in an expression, the expression is a floating point expression, and its value is a floating point number. Below are some more examples, which show that $+$, $-$ and $*$ operators can all be applied to floating point numbers, resulting in floating point numbers.

```
>>> 3.0+5.1     # floating point addition
8.1
>>> 1.0-2.4     # floating point subtraction
-1.4
>>>5.5*0.3      # floating point multiplication
1.65
```

The observation above leads to an important fact about Python: things have **types**. Literals have types. The literal 3 indicates an integer, and the literal 3.0 indicates a floating point number. Expressions have types, and their types are the types of their values. The type of a literal or expression can be examined by using the following commands:

```
>>> type(3)
<type 'int'>
>>> type(3.0)
```

```
<type 'float'>
>>> type(3+5)
<type 'int'>
>>> type(3+5.0)
<type 'float'>
```

The command *type*(x) returns the type of x. This command is a **function call** in Python, which *type* is a built-in function of Python. We call a **function** with specific **arguments** in order to obtain a specific **return value**. In the case above, calling the function *type* with the argument 3 results in the 'integer type' return value.

Function calls are also expressions, which are written in a form similar to mathematica function, with a function name followed by a comma-separated list of arguments enclosed in a pair of brackets. The value of a function call expression is the return value of the function. In the above example, *type* is the name of a function, which takes a single argument, and returns the type of the input argument. As a result, the function call *type*(3.0) evaluates to the 'float type' value.

Intuitively the return value of a function call is decided by both the function itself and the arguments of the call. To illustrate this, consider two more functions. The *int* function takes one argument and converts it into an integer, while the *float* function takes one argument and converts it into a floating point number.

```
>>> float(3)
3.0
>>> int(3.0)
3
>>> float(3)/2
1.5
>>> 3*float(3-2*5+4)**2
27.0
```

As can be seen from the examples above, when the function is *type*, the return values are different when the input argument is 3 and when the input argument is 3.0. On the other hand, when the input argument is 3, the return value of the function *type* differs from that of the function *float*. This shows that both the functions and the arguments determine the return value.

The last two examples above is a composite expression, in which the function call *float*(3) is evaluated first, before the resulting value 3.0 is combined with the literal 2 by the operator /. Function calls have higher priorities than mathematical operators in operator precedence.

The *int* function converts a floating point number into an integer by discarding all the digits after the floating point. For example,

```
>>> int(3.0)
3
>>> int(3.1)
3
>>> int(3.9)
3
```

In the last example, the return value of *int*(3.9) is 3, even though 3.9 is numerically closer to the integer 4. For floating-point conversion by rounding up an integer, the *round* function can be used.

```
>>> round(3.3)
3
>>> round(3.9)
4
```

The *round* function can round up a number not only to the decimal point, but also to a specific number of digits after the decimal point. In the latter case, two arguments must be given to the function all, with the second input argument indicating the number of digits to keep after the decimal point. The following examples illustrate this use of the *round* function with more than one input arguments. Take note of the comma that separates two input arguments.

```
>>> round(3.333, 1)
3.3
>>> round(3.333, 2)
3.33
```

A floating point operator that results in an integer value is the integer division operator (//), which discards any fractional part in the division.

```
>>> 3.0//2
1
>>> 3.5//2
1
```

Correspondingly, the modulo operator can also be applied to floating point division.

```
>>> 3.5%2
1.5
```

Another useful function is *abs*, which takes one numerical argument and returns its absolute value.

```
>>> abs(1)
1
>>> abs(1.0)
1.0
>>> abs(-5)
5
```

One final note on floating point numbers is that their literals can be expressed by a **scientific notation**. For example,

```
>>> 3e1
30.0
>>> 3e-1
0.3
>>> 3E2
300.0
```

The notations xey and $x\mathrm{E}y$ have the same meaning. They indicate the value of $x \times 10^y$.

2.1.2 Identifiers, Variables and Assignment

The set of arithmetic expressions introduced above allows simple calculations using IDLE. For example, suppose that the annual interest rate of a savings account is 4 %. To calculate the amount of money in the account after three years, with an initial sum of 3, 000 dollars is put into the account, the following expression can be used.

```
>>> 3000*1.04**3
3374.592
```

One side note is that brackets can be used to explicitly mark the intended operator precedence, even if they are redundant. In the case above, $3000 * 1.04 * *3$ can be written as $3000 * (1.04 * *3)$ to make the operator precedence more obvious. In general, being more explicit can often make the code easier to understand and less likely to contain errors, especially when there are potential ambiguities (e.g. non-intuitive or infrequently used operator precedence).

For a second example, suppose that the area of a square is $10\,\text{m}^2$. The length of each edge can be calculated by:

```
>>> 10**0.5
3.1622776601683795
```

The result can be rounded up to the second decimal place.

```
>>> round(10**0.5, 2)
3.16
```

For notational convenience and to make programs easier to maintain, Python allows names to be given to mathematical values. An equivalent way of calculating the edge length is:

```
>>> a=10
>>> w=a**0.5
>>> round(w, 2)
3.16
```

In the example above, a denotes the area of the square, and w denotes its width. The use of a and w makes it easier to understand the underlying physical meanings of the values. a and w are called **identifiers** in Python. Each Python identifiers is bound to a specific value. In the example, a is bound to 10 and w is bound to $\sqrt{10}$. Identifiers can be bound to new value:

```
>>> x=1
>>> x
1
>>> x=2
>>> x
2
```

In the example, the value of x is first 1, and then 2. Because identifiers can change their values, they are also called **variables**.

The $=$ sign in the above example is not an operator, and hence the commands $a = 10$ and $w = round(a * *0.5)$ are not expressions. They bare no values. Instead,

Table 2.1 List of keywords in Python

and	as	assert	break	class
continue	def	del	elif	else
except	exec	finally	for	from
global	if	import	in	is
lambda	not	or	pass	print
raise	return	try	while	with
yield				

the = sign denotes an *assignment* **statement**, which binds an identifier to a value. Here *a* **statement** is a command to be executed by Python, and statements are the basic execution units in Python. There are different types of statements, as will be introduced in this book. In an assignment statement, the identifier to which a value is assigned must be on the left hand side of =, and the value to assign to the identifier, which can be any expression, should be on the right hand side of =. Python gives a name to a value by binding the value to an identifier.

An intuitive difference between identifiers and literals is that the former are names while the latter are values. Formally, an identifier must start with a letter or underscore (_), and contain a sequence of letters, numbers and underscores. For example, *area*, *a*, *a0*, *area_of_square* and *_a* are all valid identifiers, but *0a*, *area of square* or *a1!* are not valid identifiers. An additional rule is that identifiers must not be **keywords** in Python, which are a list of reserved words. There are 31 keywords in total, which are listed in Table 2.1. Each keyword can be associated with one or more statements, which will be introduced in the subsequent chapters.

For another example problem, suppose that a ball is tossed up on the edge of a cliff with an initial velocity *v*0, and that the initial altitude of the ball is 0 m. The question is to find the vertical position of the ball at a certain number of seconds *t* after the toss. If the initial velocity of 5 m/s and the time is 0.1 s, the altitude can be calculated by:

```
>>> v0=5
>>> g=9.81
>>> t=0.1
>>> h = v0*t-0.5*g*t**2
>>> round(h, 2)
0.45
```

To further obtain the vertical location of the ball at 1 s, only *t* and *h* need to be modified.

```
>>> t=1
>>> h=v0*t-0.5*g*t**2
>>> round(h, 2)
0.09
```

Note that the value of *h* must be calculated again after the value of *t* changes. This is because an assignment statement binds an identifier to a value, rather than establishing a mathematical correlation between a set of variables. When $h = v0 * t - 0.5 * g * g * t ** 2$ is executed, the right hand side of = is first evaluated according to the current values of $v0$, g and t, and then the resulting value is bound to the identifier h. This is different from a mathematical equation, which establishes factual relations between values. When the value of t changes, the value of h must be recalculated using $h = v0 * t - 0.5 * g * g * t ** 2$. For another example,

```
>>> x=1
>>> x=x+1
>>> x
2
```

There are three lines of code in this example. The first is an assignment statement, binding the value 1 to the identifier x. The second is another assignment statement, which binds the value of $x + 1$ to the identifier x. When this line is executed, the right hand side of = is first evaluated, according to the current value of x. The result is 2. This value is in turn bound to the identifier x, resulting in the new value 2 for this identifier. The third line is a single expression, of which the value is displayed by IDLE. Think how absurd it would be if the second line of code is treated as a mathematical equation rather than an assignment statement!

An equivalent but perhaps less 'counter-intuitive' way of doing $x = x + 1$ is $x += 1$.

```
>>> x=1
>>> x+=1
>>> x
2
```

The same applies to $x = x - 3$, $x = x * 6$, and other arithmetic operators.

```
>>> x-=3
>>> x
-1
>>> x*=6
>>> x
-6
```

In general, *x*<*op*>=*y* is equivalent to *x*=*x*<*op*>*y*, where <*op*> can be +, −, ∗, /, % etc. The special assignment statements +=, −=, ∗=, /= and %= can be used as a concise alternative to a = assignment statement when it incrementally changes the value of one variable.

Given the fact above, it is not difficult to understand the outputs, if the following commands are entered into IDLE to find the position of the ball after 3 s in the previous problem.

```
(continued from above)

>>> t=3
>>> round(h, 2)
0.09
>>> h=v0*t-0.5*g*t**2
>>> round(h, 2)
-29.15
```

The commands above are executed sequentially and individually. When t is bound to the new value 3, h is not affected, and remains 0.09. After the new h assignment is executed, its value changes to -29.15 according to the new t. *Cascaded assignments*. Several assignment statements to the same value can be cascaded into a single line. For example, $a = 1$ and $b = 1$ can be cascaded into $a = b = 1$.

```
>>>a=b=1
>>>a
1
>>>b
1
```

2.2 The Underlying Mechanism

Floating point arithmatic can be inaccurate in calculators; the same happens in Python.

```
>>> x=1.0/7
>>> x+x+x+x+x+x+x
0.9999999999999998
>>> 3.3%2
1.2999999999999998
```

In both cases, the expression is evaluated to an imprecise number. The main reason is limitation of memory space. A floating point number can contain an infinite amount of information, if there is an infinite number of digits after the decimal point. However, the amount of information that can be processed or stored by a calculator or a computer is finite. As a result, floating point numbers cannot be stored to an arbitrary precision, and floating point operators cannot be infinitely precise. The error that results from the imprecise operations is called **rounding off error**.

At this point it is useful to know a little about the underlying mechanism of Python, so that deeper understanding can be gained on facts such as rounding off errors, which enables more solid programs to be developed. This section discusses the prepresentation and storage of numbers, the way in which arithmetic operations are carried out by Python, and the underlying mechanisms of identifiers and assignment statements.

The basic architecture of a computer is shown in Fig. 2.2 in the previous chapter. On the bottom of the figure, the main hardware components are shown, which include the CPU, memory and devices. Among the three main components, the CPU is the most important; it carries out computer instructions, including arithmetic operations. The typical way in which arithmetic operations are executed is: the CPU takes the operands from the memory, evaluates the result by applying the operator on them, and then stores the result back into the memory. The memory can be regarded as a long array of information storage units. Each unit is capable of storing a certain amount of information, and the index of its location in the long array if also referred to as its **memory address**. Devices are the channel through which computers are connected to the physical world. Keyboards, mouses, displays, speakers, microphones are a few commonly-used devices. An important device is the hard disk, on which the OS defines a file system for external storage of information.

2.2.1 Information

Computers are information-processing machines. The rounding-off error examples show that the basic unit of storage and computation can only accommodate a limited amount of information. But what is information, and how can one store information? A short answer to the first question is that, information is represented in computers as *discrete numbers*, or integers. It is an abstraction of real physical quantities, such as the pitch of sound, the colour, and the alphabet. Signals from input devices are transformed into discrete numbers before being processed by a computer, and output devices turn discrete values back into physical signals.

For example, letters typed on a keyboard are mapped into discrete numbers (e.g. 'A' → 64) before being stored into the memory. Sound waves received by a microphone are sampled at a certain rate (e.g. 256,000 times a second), and then turned into an array of discrete values. Such type of sound information can be processed (e.g. denoised) or transformed (e.g. enlarged), and than passed to a sound output device and transformed back into sound waves. A black and white display transforms a grid of numbers into a gird of pixels, each number depicting the brightness of a pixel on the display. A robot can move according to input numbers that indicate desired velocity and direction. In all these cases, devices act as a channel between computers and the physical world.

To answer the second question above, the easiest data storage medium that can be found is probably some material that can have two states (e.g. hight-voltage vs. low-voltage, solid vs. liquid, hot vs. cold). A basic storage unit made of such material can store a **binary** value, denoted as 0 or 1, each representing a distinct state of the material. Larger integers can be stored by using a combination of multiple basic storage units. For example, the combination of *two* basic storage units can store *four* distinct values: 00, 01, 10 and 11. An illustration is shown in Fig. 2.1. In general, the total number of *possible* states by combining n basic storage units is 2^n. On the other hand, at any time, the combined units can only have *one actual* combined state, which can be represented by a unique array of 0s and 1s.

It is natural to associate an array of 0 s and 1 s with a discrete number (integer). One way to number distinct states of N basic storage units $s_0, s_1, \ldots s_N$, $s_i \in \{0, 1\}$ is to interpret each distinct state as a non-negative **binary number**, treating the value of $s_N s_{N-1} \ldots s_0$ as $2^N * s_N + 2^{N-1} * s_{N-1} + \ldots + 2^0 * s_0 = \sum_{i=0}^{N} s_i \cdot 2^i$. For example, the state 101 corresponds to the integer $1 * 2^2 + 0 * 2^1 + 1 * 2^0 = 1 * 4 + 0 * 2 + 1 * 1 = 5$, and 1101 corresponds to the integer $1 * 2^3 + 1 * 2^2 + 0 * 2^1 + 1 * 2^0 = 1 * 8 + 1 * 4 + 0 * + 1 * 1 = 13$. In this way, a natural connection is established between the states of data storage materials and integers, which represent information. In computer science, each binary-valued digit is called a *bit*, and a combination of

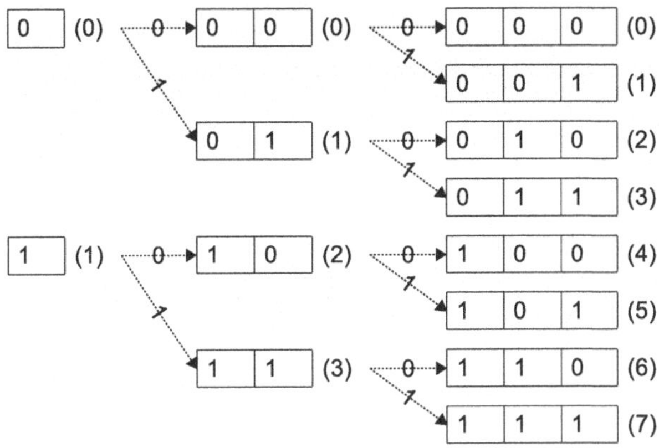

Fig. 2.1 Distinct states that can be represented by a combination of binary storage units

8 bits is a **byte**. A byte storage unit can store $2^8 = 256$ distinct numbers, which can be index by the integers 0–255. Most modern computers treat 8 bytes, or 64 bits, as a **word**, which serves as the basic units for storage.

As mentioned earlier, a digital number represents abstract information. Numbers, alphabets, sounds and all other types of information are abstracted and represented by binary numbers in computers. In **information theory**, the amount of information is measured by bits. A 64-bit word can represent 64 bits of information, which translates to 2^{64} distinct integers. The abstract information, however, can be **interpreted** in different ways. The aforementioned interpretation of information as non-negative integer values is just one example, which can be used to represent the brightness of a pixel on a black and white monitor.

Another example is the *2's complement* interpretation of **signed integers**, which uses 64 bits to represent an integer that ranges from -2^{63} to $2^{63} - 1$. In this representation, the first bit always indicates the sign: when it is 0, the number is positive; when it is 1, the number is negative. When the number is positive, the remaining 63 bits indicate the absolute value of the number. Negative numbers are represented in a slightly more complicated way. Rather than concatenating the sign bit (i.e. 1) with a 63-bit representation of the absolute value, a negative number is represented by first inversing its absolute value bit by bit, and then adding 1 to the result. For the convenience of illustration, take a byte-sized number for example. To find the representation of –13, two steps are necessary. First, the absolute value, 13, or 00001101 in binary form, is inversed bit by bit into 11110010. Then 1 is added to the back of the number:

```
    11110010
+)  00000001
-----------
    11110011
```

The resulting number, 1110011, is the binary representation of –13 in 2's complement form. Note that the first bit is 1, indicating that it is a negative number. For another example, to represent –16, its absolute value 00010000 is first inversed bit by bit into 11101111, and then 1 is added to the number,

```
   11101111
+) 00000001
-----------
   11110000
```

As a result, –13 and –16 are 11110011 and 11110000, respectively, according to 2's complement representation. The same process of negative number interpretation applies to 64-bit words. The advantage of 2's complement representations is that the addition operation between numbers can be performed in the same way regardless of whether negative numbers are involved or not. For example, $-13 + 13$ can be performed by

```
   11110011
+) 00001101
-----------
  100000000
```

With the first bit being discarded (it runs out of the 8-bit boundary, and hence cannot be recorded by 8-bit physical media), the result is 0, the correct answer.

A third useful interpretation of information is **floating point numbers**. When sound is concerned, floating point numbers give a more convenient model of the pitches in sound wave samples. As discussed earlier, all floating point numbers cannot be represented using a finite amount of information, and therefore some have to be truncated when represented using a computer word. A standard approach of representing floating point numbers in a finite number of bits is to split the total number of bits into two parts, one representing a base number (also referred to as the *significant*) and the other representing the exponent. For example, from a 64-bit word, 11 bits can be used to denote the exponent (e) and 53 bit the base (b). This type of representation naturally corresponds to the scientific notation of float literals in Python, where $bEe = b \times 10^e$. Of course, abstract information can also be interpreted as letters and other quantities in the physical world, which are out of the scope of this book.

Words are used not only as the basic units of **data storage**, but also as the basic units of **computation**. In digital circuits, of which all modern computers are made, electric signals are represented by binary values, with a high voltage in a wire denoting the value 1, and a low voltage denoting 0. Digital chips, such as CPUs, takes a fixed number of binary signals as input, and have a fixed number of output. 64-bit CPUs perform arithmetic operations on 64-bit operands, yielding 64-bit results by hardware computation. As a consequence, floating point numbers can contain only 64 bits of information, and floating point arithmetics have rounding off errors. In fact, integers are also represented by words on computer hardware, typically in 2's complement form. However, Python provides a new type, *long*, which represents numbers that exceeds the range of 64 bits. Python converts large integers into the *long* type automatically, and performs arithmetic operations between *long* type numbers implicitly by using a sequence of 64-bit integer arithmetic operations, so that programmers can use a large integer in Python without noticing the difference between *int* and *long*. *Long* type numbers can be identified by examining their types explicitly.

```
>>> type(111)
<type 'int'>
>>> type(2**64)
<type 'long'>
```

A *long* type number can also be specified explicitly using *long* literals, which are integer literals with a 'l' or 'L' added to the end

```
>>> type(111L)
<type 'long'>
```

In summary, information that a computer stores and processes is ultimately represented by a finite number of 0s and 1s (i.e. bits), organized in basic unites (e.g. words). They are interpreted in different ways when turned into specific types.

2.2.2 Python Memory Management

When evaluating an arithmetic expression, Python first constructs an **object** in the memory for each literal in the expression, and then applies the operators one by one to obtain intermediate and final values. All the intermediate and final values are stored in the memory as objects. Python objects are one of the most important concepts in understanding the underlying mechanism of Python. Integers, floating point numbers and instances of many more types to be introduced in this book, are maintained in the memory as Python objects.

Fig. 2.2 illustrates how the memory changes when some arithmetic expressions are evaluated. After IDLE starts, the memory contains some default objects, which are not under concern at this stage, and therefore not shown in the figure. When the expression 3.0 is evaluated, a new *float* object is constructed in the memory. Python always creates a new object when it evaluates a literal. When the expression $3+5*2$ is evaluated, the integer constants 3, 5 and 2 are constructed in the memory, before the operators $*$ and $+$ are executed in their precedence. When $*$ is executed, Python passes the values of the objects 5 and 2 to the CPU, together with the $*$ operator, and stores the result 10 as a new integer object in the memory. When $+$ is executed, Python invokes the CPU addition operation with the values of the objects 3 and 10, storing the result 13 as a new object.

Identifiers are names of objects in memory, used by Python to access the corresponding objects. Python associates identifiers to their corresponding values, or objects, by using a **binding table**, which is a lookup table.

Figure 2.3 shows an example of the binding table, and how it changes as Python executes assignment statements. After IDLE starts, some default entries are put into the binding table, which are ignored in this figure. When $x = 6$ is executed, the expression 6 is first evaluated, resulting in the integer object 6 in the memory. Then Python adds an entry in the binding table, associating the name x with the object 6.

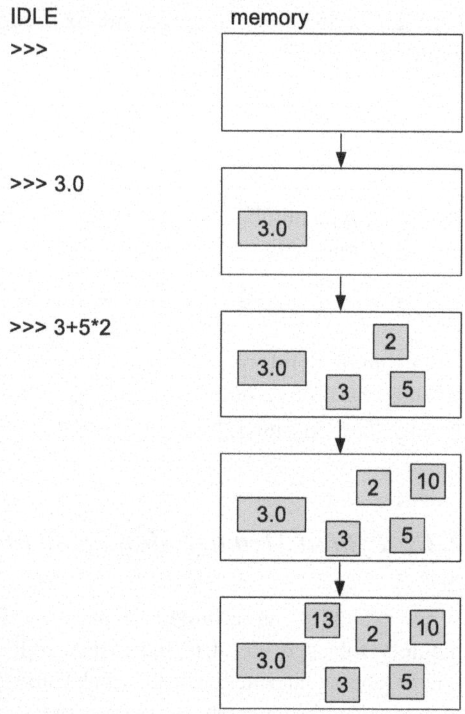

Fig. 2.2 Example memory structure for expressions

When $y = x**2$ is executed, the value of the expression $x**2$ is first evaluated by evaluating x, and 2, and then calculating $x**2$. When Python evaluates the literal 2, it creates a new object in the memory; then when it evaluates the identifier x, it looks up the binding table for an entry named x, which is bound to the object 6. The final value of the expression $x**2$ is saved to a memory object 36, and bound to the identifier y.

When the statement $x = 3$ is executed next, the expression 3 is first evaluated, resulting in a new object 3 in the memory, which is bound to the name x in the binding table. The old association between the name x and the object 6 is deleted, since one name can be bound to only one object. Note that the assignment statement *always* binds a name in the binding table with an object in the memory. Like all other Python statements, the execution is rather mechanic. Given this fact, it is easy to understand the reason why the value of y does not change to 9 automatically when $x = 3$ is executed.

At this stage, the objects 2 and 6 are still in the memory, although they are not bound to any identifiers. As the number of statements increases, the memory can be filled with many such objects. They are no longer used, but still occupy memory space. Python has a **garbage collector** that periodically removes unused objects from the memory, so that more memory space can be available. Note also that Fig. 2.2 does

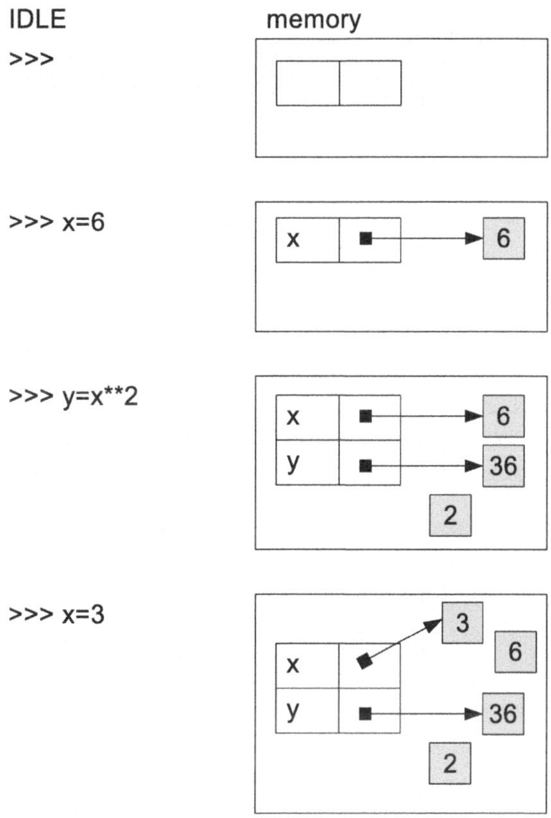

Fig. 2.3 Example memory structure for assignments

not show the binding table, although it exists in the memory, containing identifiers that are irrelevant to the example.

For readers who know C++ and Java variables, it is worth nothing that Python variables are not exactly the same as their C++ and java counterparts. Specifically, the assignment statement in Python changes the value of a variable by changing the binding (i.e. associating the identifier to a different object in the binding table), while the assignment statement of C++ and Java directly changes the value of the object that the identifier is associated with, without changing the binding between identifiers and memory objects. Although in many cases, Python variable can be used in the same way as C++ and Java variables from the programming perspective, an understanding of this difference could be useful in a voiding subtle errors, especially when mutability (Chap. 7) is involved.

Python provides a special statement, the **del statement**, for deleting an identifier from the binding table.

```
>>> x=1
>>> y=2
>>> x
```

```
1
>>> y
2
>>> del x
>>> x
Traceback (most recent call last):
  File "<pyshell#5>", line 1, in <module>
    x
NameError: name 'x' is not defined
>>> y
2
```

As can be seen from the example above, the *del* statement begins with the *del* key word, followed by an identifier. It deletes the identifier from the binding table. As the second last command shows, Python reports a *name error* when the value of x is requested after the identifier x has been deleted from the binding table. After an identifier is deleted, the objected that it is bounded to is not deleted immediately. Instead, the garbage collector will remove it later when no other identifiers are bound to it.

Each Python object has a unique **identify**, which is typically its memory address. Python provides a function, *id*, which takes a Python object as its input argument, and returns the identify of the object. For example,[2]

```
>>> x=12345
>>> id(x)
4535245720
>>> y=x
>>> id(y)
4535245720
>>> y=23456
>>> id(x)
4535245720
>>> id(y)
4535245672
>>> id(12345)
4535245984
```

When x is assigned to the value 12345, it is bound to a new object 12345 in the memory. When y is assigned to the value of x, it is bound to the same object 12345. Hence the identifies of x and y are the same. When y is reassigned to the value 23456, a new object 23456 is constructed in the memory, occupying a new memory address, and y is bound to his object. Hence the identify of y changes, while the identify of x remains the same. The last command shows the identify of a new object, constructed by the evaluation of the expression 12345. It is different from that of x, because every time a literal is evaluated, a new object is constructed in the memory. Here is another example.

```
>>> x=12345
>>> y=x
>>> id(x)
```

[2]The memory addresses in the examples below are unlikely to be reproduced when the examples are tried again, since the allocation of memory addresses is dynamic and runtime dependent.

```
4462893976
>>> id(y)
4462893976
>>> x=12345
>>> id(x)
4462894072
>>> id(y)
4462893976
```

When x is assigned to the value 12345 the second time, the right hand side of the assignment statement is evaluated first, which leads to a new object having the value 12345 in the memory. x is bound to this new object, while y remains the same. Although the values of x and y are the same, their memory addresses, or identifies are different, because they are bound to two different objects.

Note that the observations above may not hold for small numbers (e.g. 3 instead of 12345). This is because to avoid frequent construction of new objects, Python constructs at initialization a set of frequently used objects, including small integers, so that they are reused rather than constructed afresh when their literals are evaluated.

```
>>> x=10
>>> y=10
>>> id(x)
140621696282752
>>> id(y)
140621696282752
```

In the example above, y is assigned the value 10 after x is assigned the value 10. In each case, no new object is created when the literal 10 is evaluated, because an object with the value 10 has been created in the memory location 140621696282752 to represent all objects with this value.[3]

In summary, in addition to a value, a Python object also has an identify and a type. Once constructed, the identify and type of an object cannot be changed. For number objects, the value also cannot be changed after the object is constructed.

2.3 More Mathematical Functions Using the *math* and *cmath* Modules

Several mathematical functions have been introduced so far, which include addition, subtraction, multiplication, division, modulo and power. There are more mathematical functions that a typical calculator can do, such as factorial, logarithm and trigonometric functions. These functions are supported by Python through a special **module** called *math*.

A Python module is a set of Python code that typically includes the definition of specific variables and functions. The next chapter will show that a Python module can

[3]Exactly which small values are represented as frequently accessed objects depends on the Python distribution.

be nothing but a normal Python program. In order to use the variables and functions defined in a Python module, the module must be *imported*. For example,

```
>>> import math
>>> math.pi
3.141592653589793
>>> math.e
2.718281828459045
```

In the example above, the *math* module is imported by the statement *import math*. The *import* statement is the third type of statement introduced in this chapter, with the previous two being the assignment statement and the *del* statement. The *import* statement loads the content of a specific module, and adds the name of the module in the binding table, so that content of the module can be accessed by using the name, followed by a dot (.).

Two mathematical constants *pi* and *e*, are defined in the *math* module, and accessed by using '*math.*' in the above example. Both *pi* (π) and *e* are defined as floating point numbers, up to the precision supported by a computer word.

Mathematical functions can be accessed in the same way as constants, by using '*math.*'. For example, the *factorial* function returns the factorial of the input argument.

```
>>> import math
>>> math.factorial(3)
6
>>> math.factorial(8)
40320
```

The *math* module provides several classes of functions, including power and logarithmic functions, trigonometric functions and hyperbolic functions. The two basic power and logarithmic functions are *math.pow*(x, y) and *math.log*(x, y), which take two floating point arguments x, and y, and return x^y and $\log_y x$, respectively.

```
>>> import math
>>> math.pow(2, 5)
32.0
>>> math.pow(25,0.5)
5.0
>>> math.log(1000,10)
2.9999999999999996
```

Note the rounding off error in the last example. The functions *math.pow* and *math.log* always return floating point numbers. The function *math.log* can also take one argument only, in which case it returns the natural logarithm of the input argument.

```
>>> import math
>>> math.log(1)
0.0
>>> math.log(math.e)
1.0
>>> math.log(10)
2.302585092994046
```

There is also a handy function to calculate the base-10 logarithm of an input floating point number: *math.log10*(x). The function take a single floating point argument. As two other special power functions, *math.exp*(x) can be used to calculate the value of e^x, and *math.sqrt*(x) can be used to calculate the square root of x.

The set of trigonometric functions that the *math* module provide include *math.sin* (x), *math.cos*(x), *math.tan*(x), *math.asin*(x), *math.acos*(x) and *math.atan*(x), which calculate the sine, the cosine, the tangent, the arc sine, the arc cosine, the arc tangent of x, respectively.

```
>>> import math
>>> math.sin(3)
0.1411200080598672
>>> math.cos(math.pi)
-1.0
>>> math.tan(3*math.pi)
-3.6739403974420594e-16
>>> math.asin(1)
1.5707963267948966
>>> math.acos(1)
0.0
>>> math.atan(100)
1.5607966601082315
```

Note the rounding off errors in some of the examples above. All angles in the functions above are represented by radians. The *math* module provides two functions to convert between radians and degrees: the *math.degrees* function takes a single argument x, and converts x from radians to degrees; the *math.radians* function takes a single argument x, and converts x from degrees to radians.

The set of hyperbolic functions include *math.sinh*(x), *math.cosh*(x), *math.tanh*(x), *math.asinh*(x), *math.acosh*(x) and *math.atanh*(x), which calculate the hyperbolic sine, the hyperbolic cosine, the hyperbolic tangent, the inverse hyperbolic sine, the inverse hyperbolic cosine, and the inverse hyperbolic tangent of x, respectively.

There are more functions that the *math* module provides, including *math.ceil*(x), which returns the smallest integer that is greater than or equal to x, and *math.floor*(x), which returns the largest integer that is less than or equal to x. It does not make sense to remember all the functions that Python provides for the purpose of programming, although remembering a few commonly-used functions would be useful for the efficiency of programming. A good practice is to keep the **Python documentation** at hand, which is also easily accessible online. For example, searching for the key words 'Python math' using a search engine can lead to the Python documentation for the *math* module.

2.3.1 Complex Numbers and the cmath *Module*

Complex literals. In some engineering disciplines, *complex numbers* are commonly useful. A complex number consists of a *real* part and an *imaginary* part. While the real part of a complex number is an arbitrary real number, represented by a floating point number in Python, the imaginary part is based on j, the imaginary square root of -1. Python represents the imaginary part of a complex literal by a floating point number, followed by the special character j.

```
>>> c=1j
>>> type(c)
<type 'complex'>
>>> c*c
(-1+0j)
```

In the example above, c is a complex number that has only the imaginary part, $1j$. The square of c is $1j \times 1j = -1$. In Python, if a complex number co-exists in an expression with floating point numbers and integers, the type of the whole expression becomes a complex number. Therefore, the value of the expression $c * c$ is the complex number $(-1 + 0j)$.

$1j$ is a special complex number, with the real part being 0. In general, a complex number is specified by the sum of its real part and imaginary part, as shown by IDLE in the example above.

```
>>> a=1+2j
>>> a
(1+2j)
>>> type(a)
<type 'complex'>
>>> b=3+4j
>>> b
(3+4j)
>>> type(b)
<type 'complex'>
```

Type conversion from integers and floating point number into complex numbers. Similar to the construction of integer and floating number objects using the functions *int* and *float* introduced earlier, a complex number can be constructed from integers and floating point numbers by using the function *complex*.

```
>>> complex(1,2)
(1+2j)
>>> a=complex(-1, 0.5)
>>> a
(-1+0.5j)
```

As shown by the example above, the function *complex* takes two numeric arguments specifying the real and imaginary components, respectively, and returns a complex object.

Complex operators. Similar to integers and floating point numbers, complex numbers also support arithmetic operations by using operators. The $+$, $-$, $*$, $/$ and $**$ operators for integers and floating point numbers also apply to complex numbers.

```
>>> a=1+2j
>>> b=-1+3j
>>> a+b
5j
>>> a-b
(2-1j)
```

```
>>> a*b
(-7+1j)
>>> a/b
(0.5-0.49999999999999994j)
>>> a**2
(-3+4j)
```

Functions for complex numbers. The *abs* function, when applied to complex numbers, returns the magnitude of the number.

```
>>> a=3+4j
>>> abs(a)
5.0
```

In the example above, the input argument to the function call *abs*(a) is a complex number, and the return value is a floating point number. However, *no* built-in function or operator takes floating point numbers but results in a complex number. In other words, the default domain in which Python handles mathematical expressions is real numbers. For example, trying to obtain the square root of -1 by the $**$ operator, or using the *math.sqrt* function, will result in an error.

```
>>> (-1)**0.5
Traceback (most recent call last):
  File "<stdin>", line 1, in <module>
ValueError: negative number cannot be raised to a
    fractional power
>>> import math
>>> math.sqrt(-1)
Traceback (most recent call last):
  File "<stdin>", line 1, in <module>
ValueError: math domain error
```

This choice of the default mathematical domain for Python is based on the fact that real numbers are the most widely used, while complex numbers are common only in specific fields. Some users of Python might not even know the existence of complex numbers. As a result, the most natural choice is to leave their processing only to specific modules. Python provides a module, *cmath*, for complex numbers.

The *cmath* module. The power and logarithmic functions that the *cmath* module provides include *cmath.exp*, *cmath.log*, *cmath.log10* and *cmath.sqrt*. They bare the same names as their counterparts in the *math* module, with the difference being that they can be applied to complex numbers, and can return complex numbers. For example, to get the square root of -1 in the complex domain, the *cmath.sqrt* function should be used.

```
>>> import cmath
>>> cmath.sqrt(-1)
1j
```

The *cmath* module also provides the trigonometric functions *cmath.sin*, *cmath.cos*, *cmath.tan*, *cmath.asin*, *cmath.acos* and *cmath.atan*, and the hyperbolic functions *cmath.sinh*, *cmath.cosh*, *cmath.tanh*, *cmath.asinh*, *cmath.acosh* and *cmath.atanh*, with exactly the same use as their *math* counterparts except of the domain.

While the *abs*(x) function returns the magnitude of a complex number x, the function *cmath.phase*(x) returns the phase of x. The *cmath.polar*(x) function returns the representation of a complex number x in polar coordinates, while the function *cmath.rect*(r, p) returns the complex number x given it polar coordinates (r, p).

2.3.2 Random Numbers and the random Module

For one last example of modules in this chapter, the *random* module provides functions for generating random numbers. It is useful to a range of mathematical problems, including a branch of numerical simulation methods that will be introduced in this book.

Two important functions provided by the *random* module include *random.random* and *random.randint*. *random.random*() takes no input arguments, and returns a random floating point number in the range [0.0, 1.0). *random.randint*(a, b) takes two integer input arguments a and b, and returns a random number between a and b, inclusive.

```
>>> import random
>>> random.random()
0.1265646141812915
>>> random.random()
0.5701294637390362
>>> random.random()
0.8842027581970576
>>> random.randint(10,20)
12
>>> random.randint(10,20)
17
>>> random.randint(10,20)
17
>>> random.randint(10,20)
19
```

In general, many commonly-used mathematical functions are provided by Python, and it would always be useful to look for a readily-available implementation via the Python documentation and other resources. However, there are also cases where a customized function is needed. The following chapters will introduce step by step how complex functionalities can be achieved by the powerful Python language.

Exercises

1. What are the values of the following expressions?

   (a) $1 + 3 * 2 - 5 + 4$
   (b) $1 + 3 * (2 - 5) + 4$
   (c) 5**2**2*3+1
   (d) 5**(2**2)*3+1
   (e) 1+3/2

(f) 1+3.0/2
(g) −2−1
(h) −(2−1)
(i) 3.0 + 3/2
(j) 3 + 3/2.0
(k) −1**0.5

2. Use IDLE to calculate the following mathematical values.

   (a) 10^5
   (b) $\sqrt{10}$
   (c) the roots of $x^2 - 7x + 10 = 0$
   (d) $lg(2 + \sqrt{5})$
   (e) the area of a circle with a radius of 5.5
   (f) $sin2.5$
   (g) the complex roots of $x^2 - 2x + 10 = 0$
   (h) 4!
   (i) $\sum_{k=32}^{128} k$
   (j) $\prod_{k=3}^{17} k$

3. What are the values of the following binary numbers if they are (a) non-negative; or (b) 2's Complements?

   (a) 01001
   (b) 100
   (c) 1100
   (d) 11111
   (c) 11111111

4. Use IDLE to solve the following mathematical problems.

   (a) A car runs at a constant speed of 20 km/h. When it passes another case, the latter starts to accelerate in order to catch up. Assuming that the first case keeps a constant speed, and the second car keeps a constant acceleration of $2\,m/s^2$. After how many seconds will the second car catch up with the first one?
   (b) The annual interest rate of a savings account is 4.1 %. John has $10,000 in his account, and aims at saving $50,000 within 5 years by depositing a fixed amount of money to his account in the beginning of each year, including this year. How much money does John need to save each year in order to achieve his goal?
   (c) In a shooting exercise, a coach stands 5 m away from a trainee, and throws a target up vertically at 5 m/s. If the trainee must fire her gun exactly 0.5 s after the throwing of the ball, then at which angle should she aim? If she must fire the gun exactly 1 s after the throwing, then at which angle should she aim?

(d) John deposited an initial sum of $3000 in his account. After 3 years, the balance reaches $3335.8 due to composite interest. What is the interest rate per annual?
(e) The three sides of a triangle are 3, 4 and 6m, respectively. What is its area?

5. What are the three most important properties of a Python object? Which of them can change after the object is constructed, if the object is a number?
6. State the main differences between identifiers and literals. Given a token in a program, how does Python know whether it is an identifier or a literal?

Chapter 3
The First Python Program

This chapter shows how to turn the computation power of the previous chapter into stand-alone Python programs, which can be executed on their own, independent of IDLE. To achieve such independence, the most important additional functionality that is required is input and output—or communication with users.

3.1 Text Input and Output Using Strings

Text input and output (I/O) used to be the dominant user interface before graphical user interface (GUI) was invented in the 1980s. Nowadays, most system administrators, researchers and hackers still use the text console as the main I/O interface. Compared with GUI, text UI is relatively much easier to learn as a beginner. Python allows interaction between a program and a user via text IO, by proving a special type of objects: strings. A Python *string* can be treated as a sequence of text characters.

String literals can be written in three different formats, the first being letters wrapped between a pair of single quotes:

```
>>> s='abc'
>>> s
'abc'
>>> type(s)
<type 'str'>
```

In the example above, there are three text characters in the string *s*, namely 'a', 'b' and 'c'. The type of the string *s* is *str*, which is the Python representation of the string type.

The second way to write a string literal is to replace single quotes with double quotes; this is convenient if the string itself contains single quotation marks.

```
>>> s="he said:'hello'"
>>> s
"he said:'hello'"
>>> type(s)
<type 'str'>
```


Y. Zhang, *An Introduction to Python and Computer Programming*,
Lecture Notes in Electrical Engineering 353, DOI 10.1007/978-981-287-609-6_3

In case both single and double quotes are in a string itself, a third form of string literals can be used. It is specified by putting three double quotes on the left hand side and three on the right hand side of the string. This form of string literals also allows a string to span over multiple lines:

```
>>> s="""abc
... def
... ghi
..."""
>>> s
'abc\ndef\nghi\n'
>>> type(s)
<type 'str'>
```

Line breaks are represented by a special character in the string, which can be written with the **escaped form** '\\n'. By default, Python displays the value of a string object in its escaped format, which is demonstrated by the example above. Escaped characters can also be used in a string literal directly.

```
>>> s="abc\ndef\nghi"
>>> s
'abc\ndef\nghi'
```

In the example above, the two-character sequence '\\$n$', is used in the literal explicitly to represent a new-line character. Note that explicit (i.e. not escaped) line breaks are not allowed in string literals with single or double quotes. Another escape character that has an escape form is the tab character, which can be written as '\\$t$'.

```
>>> s='abc	def	ghi'
>>> s
'abc\tdef\tghi'
```

In the example above, the string s contains two explicit tab characters, which are shown in escaped forms by Python. Note that the quotation marks around strings serve only as indicators of a string literal—they are not a part of the string itself. Hence '' or "" specifies an empty string, a string that does not contain any character.

String operators. Similar to integers and floating point numbers, string objects also support a set of operators. A commonly used string operator is the *concatenation* operator +:

```
>>> s1='abc'
>>> s2='def'
>>> s1+s2
'abcdef'
```

Figure 3.1 illustrates the changes that happen to the memory when the lines of code above are executed. When the string literals '*abc*' and '*def*' are evaluated, two corresponding string objects are constructed in the memory. They are associated with their respective identifiers via the assignment statements. When the concatenation operator is applied, a new string object is constructed, taking the concatenated value of the two operands.

One thing to note is that the same operator + behaves differently when applied to strings as compared to numbers. This fact is an example of **polymorphism**, which will be discussed in Chap. 10.

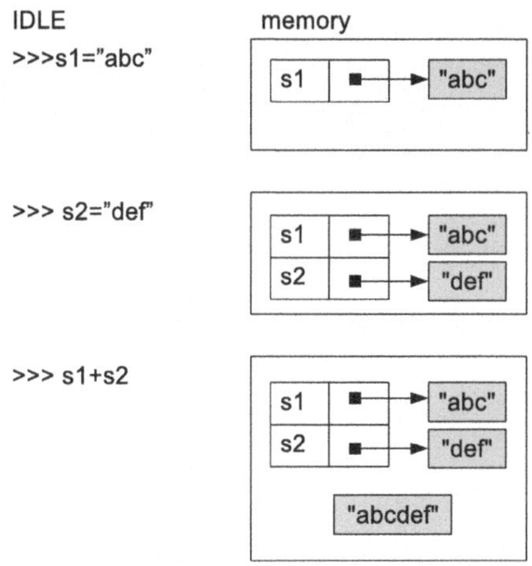

Fig. 3.1 Example memory structure for string operations

The $*$ operator can also be applied to strings. It takes a string operand and an integer operand, resulting in a string by repeating the string operand a number of times, as specified by the integer operand.

```
>>> s="abc"
>>> t=s*3
>>> t
'abcabcabc'
```

One important operator of strings is the *getitem* operator([]), which takes a string operand and an integer operand, resulting in the character in the string at the index specified by the integer. The *getitem* operator does not apply to numbers, such as integers and floating point numbers, but applies to other *sequential* objects, which will be introduced later in Chaps. 5 and 6. It takes the form of a pair of squared brackets, written immediately after the string operand and enclosing the integer operand, which specifies for the character index.

```
>>> s='abc'
>>> s[0] # the first character
'a'
>>> s[1] # the second character
'b'
>>> s[2]
'c'
>>> x=0
>>> s[x] # index can be any integer expression
'a'
```

Note that string indices start from 0 rather than 1. Hence $s[i]$ stands for the $i+1$th character in s when $i \geq 0$. This is also true for other sequential types in Python, and sequential types in many other programming languages.

Negative numbers can also be used to specify indices in Python. Starting from -1, negative indices specify character indices from the right of the string.

```
>>> s='abc'
>>> i=-1
>>> s[i]
'c'
>>> i-=1
>>> i
-2
>>> s[i]
'b'
>>> i-=1
>>> i
-3
>>> s[i]
'a'
```

String indices must be within the valid range—trying to get a character that does not exist in the string will result in an error.

```
>>> s='abc'
>>> s[3]
Traceback (most recent call last):
  File "<stdin>", line 1, in <module>
IndexError: string index out of range
>>> s[-4]
Traceback (most recent call last):
  File "<stdin>", line 1, in <module>
IndexError: string index out of range
```

A **sub string** can be extracted from a string by using the *getslice* operation ([]), which similar to the *getitem* operation by using the [] operator. It is expressed by replacing the character index in a *getitem* expression with a *slice*, which consists of a start and an end index, separated by a colon (:)

```
>>> s='abc'
>>> s[1:2]   # indices 1
'b'
>>> s[0:2]   # indices 0,1
'ab'
>>> s[2:3]   # indices 2
'c'
```

The way in which *getslice* is performed is illustrated by Fig. 3.2, where the locations at which a slice is taken is worth noting. Both the start index and the end index specify positions *before* characters. Similar to *getitem*, character indices start from 0. For example, '$s[0 : 2]$' takes a slice out of s, by starting from *before* the first character (index = 0), and ending *before* the third character (index = 2), resulting in consisting of the first and second characters. Unlike *getitem*, *getslice* will not result in an index error; if the ending index is out of range, the slice will end with the last character in the string.

```
>>> s='abc'
>>> s[2:5]
'c'
>>> s[-2:-1]
'b'
```

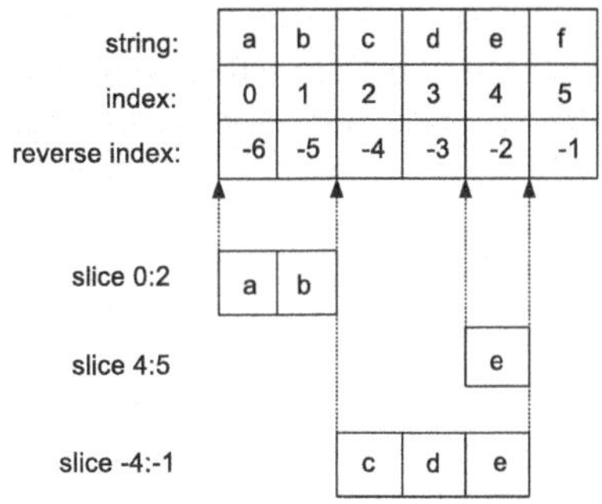

Fig. 3.2 Illustration of getslice

By making the end index smaller than or equal to the start index, an empty string results from *getslice*.

```
>>> s='abc'
>>> s[2:2]
''
>>> s[2:1]
''
```

One or both indices in a *getslice* operation can be unspecified. The default value for an unspecified start index is 0, while the default value for an unspecified end index is the end of string. For example,

```
>>> s='abc'
>>> s[1:]    # default end
'bc'
>>> s[:2]    # default start
'ab'
>>> s[:]    # both default
'abc'
```

To enhance the flexibility of slicing, a slice can be further specified by adding a third parameter, *step*, which is joined to the start and end indices by using an additional colon (:), and indicates an interval at which characters are taken from the string. For example,

```
>>> s='abcdef'
>>> s[0:6:2] # 0, 2, 4
'ace'
>>> s[0:6:3]
'ad'
>>> s[1:5:2]
'bd'
```

When the step is 2, every second character is taken from the string; when the step is 3, every third character is taken. The start and end indices specify the same locations with or without the step parameter. The default value of the step parameter is 1, which results in a continuous sub string. When the step is larger than 1, a discontinuous sub string is extracted.

The step parameter can also be negative, in which case the slice is taken from right to left (i.e. **right-to-left slicing**), and hence the start index must be larger than the end index. Different from the left-to-right slicing, when slicing from right to left, indices indicate locations *after* corresponding characters. Similar to the left-to-right slicing, the absolute value of the step parameter specifies the interval at which characters are taken. For example,

```
>>> s='abcdef'
>>> s[5:2:-1]    # 5, 4, 3
'fed'
>>> s[:2:-1]
'fed'
>>> s[::-1]
'fedcba'
>>> s[::-2]
'fdb'
```

An important function that is associated with strings is *len*, which takes a single string argument and returns the number of characters in the string.

```
>>> s='abc'
>>> len(s)
3
>>> s='abcdef'
>>> len(s)
6
>>> s=''
>>> len(s)
0
```

As shown in the last example above, the length of an empty string is 0.

Conversion between strings and other types. Similar to the *int* and *float* functions for type conversion into integers and floating point numbers, the *str* function can be used to convert floating point numbers and integers into strings.

```
>>> i=123
>>> f=1.23
>>> s1=str(i)
>>> s1
'123'
>>> type(s1)
<type 'str'>
>>> s2=str(f)
>>> s2
'1.23'
>>> type(s2)
<type 'str'>
```

In the opposite direction, the *int* and *float* functions can turn strings that represent integers and floating point numbers to integer and floating point objects, respectively.

```
>>> s='123'
>>> int(s)
123
>>> s='1.23'
>>> float(s)
```

```
1.23
>>> int(s)
Traceback (most recent call last):
  File "<stdin>", line 1, in <module>
ValueError: invalid literal for int() with base 10: '1.23'
>>> s='abc'
>>> float(s)
Traceback (most recent call last):
  File "<stdin>", line 1, in <module>
ValueError: could not convert string to float: abc
```

The conversion from strings to integers and floating point numbers is strict: when the string does not correspond to the target type, an error occurs. Two special strings that can be converted to a floating point numbers are 'inf' and '-inf', which represent infinity and negative infinity, respectively. These two special floating point numbers do not have literals, and must be converted from the strings 'inf' and '-inf'.

```
>>> a=float('inf')
>>> type(a)
<type 'float'>
>>> a
inf
>>> b=float('-inf')
>>> type(b)
<type 'float'>
>>> b
-inf
```

When a floating point number is converted into a string, Python automatically rounds it up to a certain number of digits after the decimal point, so that the output is more readable.

```
>>> 1.0/7
0.142857142857
>>> s3=str(1.0/7)
>>> s3
'0.142857142857'
>>> type(s3)
<type 'str'>
```

If a specific number of digits after the decimal point is required in the output string, the *round* function introduced in the previous chapter can be used to round up the floating point number before it is converted into a string. Alternatively, a **string formatting** expression can be used. String formatting expressions are a powerful tool for specifying the format of numbers in a string. As a first example, consider the formatting of integers:

```
>>> s='%d'%123 # normal form
>>> s
>>> '123'
>>> s='%5d'%123 # padding to the left
>>> s
'  123'
>>> s='%-5d'%123 # padding to the first
>>> s
```

```
'123   '
>>> s='%1d'%123 # no padding
>>> s
'123'
```

A string formatting expression consists of two parts, separated by a % symbol (e.g. *'%5d'%123*). The part on the left is a **pattern string** that contains a **pattern** *%xd* (e.g. *'%5d'*), where % indicates the start of a pattern, and the letter *d* indicates that the formatted pattern is an integer. On the right hand side of the % symbol is the **argument** to fill the pattern, and in this case it is an integer to be formatted in the string (e.g. 123). The *x* in the pattern is optional; it indicates the length of the string: if it is positive, spaces are padded on the left when the integer contains less digits than *x*, and if it is negative, spaces are padded on the right. If the integer contains more digits than the size specified by *x*, no space will be padded but the integer will not be truncated either.

In addition to patterns, the pattern string can consist of other characters. Characters that are not a part of a pattern will remain unchanged when patterns are replaced with arguments during string formatting.

```
>>> '%dabc'%123 # pattern = %d
'123abc'
>>> '%5dabc'%123 # pattern = %5d
'  123abc'
>>> 'abc%-5dd'%123 # pattern = %-5d
'abc123 d'
```

To format a floating point number, the pattern in a pattern string is *%x.yf*, where *x* specifies the total size of the string, in the same way as integer formatting, *y* specifies the number of digits after the decimal point, and *f* marks a floating point pattern. *x* can be omitted, in which case no padding will be added.

```
>>> '%f'%1.23 # no padding
'1.23'
>>> '%5.2f'%1.23 # padding to the left; two digits
    after the decimal point
' 1.23'
>>> 'abc%5.2f'%1.23
'abc 1.23'
>>> 'abc%5.2f'%(1.0/7)
'abc 0.14'
>>>'abc%.2f'%(1.0/7) # no padding
'abc0.14'
```

More than one patterns can be defined in the pattern string, in which case a comma-separated list of arguments must be given on the right within a pair of brackets. The patterns will be filled by the arguments in their input order.

```
>>> 'a string with one integer %d followed by one
    floating point number %.2f' % (123, 1.23)
'a string with one integer 123 followed by one
    floating point number 1.23'
```

If the number of patterns does not match the number of arguments, an error will be given:

```
>>> '%d %d %f' % (1, 2)
Traceback (most recent call last):
  File "<stdin>", line 1, in <module>
TypeError: not enough arguments for format string
```

String representation in memory. As mentioned in the previous chapter, all types of information, including the alphabet, are ultimately represented by abstract information, or binary numbers, in memory. Python provides a built-in function, *ord*, which takes a single character as the only argument, and returns the integer that represents (i.e. encodes) the character:

```
>>> ord('a')
97
>>> ord('b')
98
>>> ord('z')
122
>>> ord('A')
65
>>> ord('B')
66
>>> ord('Z')
90
```

As can be seen from the example above, the representations of 'A'—'Z' are 65–90, and the representations of 'a'—'z' are 97–122.

The reverse function to *ord* is *chr*, which takes an integer between 0 and 255 as the only argument, and returns the character interpretation of the integer.

```
>>> chr(97)
'a'
>>> chr(98)
'b'
>>> chr(106)
'j'
```

3.1.1 Text IO

The print statement is used for text output; it shows the value of an object as a string on the *text console*. In most cases, the text console is the text terminal (i.e. *Terminal* in Linux and Mac OS, or *Command Line* in Windows). In the case of IDLE, the text console is IDLE itself. The syntax of the *print* statement is *print x*, where *x* is an expression. When *print x* is executed, the value of *x* is calculated first, and then converted into a string, before being displayed.

```
>>> print 123
123
>>> print 3.6
3.6
>>> print (1.0/7)
0.142857142857
>>> print 'abc'
abc
>>> print 3+5-2*6
-4
>>> print '%.2f'%3.1415
3.14
```

In the last two examples above, the values of the expressions $3 + 5 - 2 * 6$ and $'\%.2f'\%3.1415$ are evaluated first, before being displayed on the console, respectively. Here the string type is used as a bridge between the value of an expression and the output console—an object is converted into a string before being displayed on the console.

```
>>> 1.0/7
0.14285714285714285
>>> str(1.0/7)
'0.142857142857'
>>> print 1.0/7
0.142857142857
```

In the example above, the first two lines of code show Python's internal representation of the internal representation of the floating point number and its string conversion, respectively. The third line of code is a *print* statement that writes the value of the floating point number on the console. As can be seen from the example, the floating point number displayed by the *print* statement is the same as the string conversion, but different from the internal representation of the floating point number itself. This is because the number has gone through a string conversion before being printed.

The string conversion by the *print* statement is *implicit*—the *str* function is not called explicitly in the statement. Later in Chap. 10 when *custom types* are introduced, a custom string conversion process is discussed, and the implicit string conversion by the *print* statement is reflected more directly.

The *print* statement displays a string on the console. This may appear to be a redundant functionality, since IDLE will display the value of an expression anyway, as the first two lines of code in the previous example demonstrate. However, as will be shown in the next section, the *print* statement is necessary and important when Python code is executed as a stand-alone program and independent of IDLE. In that case, the value of an expression is *not* displayed when the expression is a single line of code. In addition, although for integers and floating point numbers, Python's internal representation is *similar* to the string conversion, for many other types, the two representations can be very different (see more details in Chap. 11), and hence what IDLE displays can be rather different from what the *print* statement shows.

One final note is that when a string is printed, the escaped characters in the string are displayed in their original form.

```
>>> s='abc\ndef\nghi\tj'
>>> s    # internal representation
'abc\ndef\nghi\tj'
>>> print s # print version
abc
def
ghi      j
```

Because backslach(\) are used for escaped characters, their literal form in a string is represented by '\\'.

```
>>> s = 'backslach is \\'
>>> s
'backslach is \\'
>>> print s
backslach is \
```

Input from the text console can be achieved by using the *raw_input* function, which asks the user to enter a string and returns the content of the string as a Python object. A command line prompt can be specified as an input argument to this function.

```
>>> s=raw_input("Enter a string:")
Enter a string: abc
>>> s
'abc'
```

In this example, when the first line of code is executed, the input argument "Enter a string:"is displayed, with the program being temporarily stopped to wait for user input. The user enters the letters 'a', 'b' and 'c', and then presses the *Enter* key to complete the input. The function call then returns the string "abc" as its value.

Note that different from the case of the *print* statement, *raw_input("Enter a string:")* is a function call. It forms an expression on the right hand side of an assignment statement in the example above, and is evaluated to a string object. The specialness of this function call is its evaluation process: IDLE stops to ask the user for text input, and evaluates the return value of the function call according to the input, rather than according to the input string argument (e.g. *"Enter a string:"*), which is used for console prompt only. This behavior is rather different from the *round*, *len* and *int* function calls, which are all numerical and do not affect the execution of IDLE. However, it remains true that all function calls are expressions that evaluate to a Python object.

Being an expression itself, *raw_input* calls could be used as a part of a more complex expression. For example,

```
>>> i=int(raw_input("Enter a number:"))*2
Enter a number: 56
>>> i
112
```

When the assignment statement above is executed, the right hand side of = is first evaluated. It consists of (1) the evaluation of the *raw_input* function call, (2) the evaluation of the *int* function call, with the result of (1) being used as the input argument, and (3) the evaluation of the * operator, with the result of (2) being used as one operand and the integer object 2 as the other. In step (1), IDLE displays the prompt message and stops to ask the user for an input. The user enters the characters '5', '6', and presses *Enter*. The function therefore returns the string object '56'. In step (2), the string is converted into the integer object 56. In step (3), the object is multiplied by 2, resulting in the final value of the whole expression, 112. This object is in turn bound to the identifier *i* in the assignment.

raw_input is the basic mechanism in Python for accepting user input from a text console. The return value of this function is always a string, and therefore type conversion is necessary when the desired object is an integer, a floating point number or other types. An alternative function for text console input is *input*, which performs type conversion *implicitly*. The *input* function is implemented by adding an automatic type conversion step after a *raw_input* step. In effect, the function receives a *literal* from the user, and converts it to a Python object as the return value.

```
>>> x=input("x=")
x=123
>>> type(x)
<type 'int'>
>>> x
123
>>> x=input("x=")
x=12.3
>>> type(x)
<type 'float'>
>>> x
12.3
>>> x=input("x=")
x='abc'
>>> type(x)
<type 'str'>
>>> x
'abc'
>>> x=input("x=")
x=1+2j
>>> type(x)
<type 'complex'>
>>> x
(1+2j)
>>> x=input("x=")
x=1L
>>> type(x)
<type 'long'>
>>> x
1L
>>> x=input("x=")
x=abc
Traceback (most recent call last):
  File "<stdin>", line 1, in <module>
  File "<string>", line 1, in <module>
NameError: name 'abc' is not defined
```

In the example above, the *input* function is called with the "x=" prompt being displayed. An integer literal, a floating point number literal, a string literal, a complex number literal, an explicit long number literal and an arbitrary string are entered, respectively. While the literals of the corresponding types are converted to individual objects, the arbitrary string, *abc*, causes an error, because it does not represent any literal. In fact, it is treated as an identifier, and hence Python tries to look for the objects that is associated with the name *abc* in the binding table, leading to a *name error*. The correct way to enter a string literal, as shown by the 'abc' example, is to include the quotations.

Because of the extra layer of functionality, the *input* function offers more flexibility compared with the *raw_input* function. However, it can also cause unexpected errors if the user is not familiar with Python literals. From the programmer's perspective, the possibility for the return value of a function to be an arbitrary type can also lead to difficulty in processing it. As a result, the *raw_input* function with explicit type conversion can be preferred when a certain type of object is expected from users.

3.2 The First Python Program

Text IO allows a stand-alone Python program to interact directly with the user. For a simple Python program that does not rely on IDLE, consider the problem of interest calculation. Suppose that a user has an initial sum of money in a savings account, and the interest rate is fixed for a number of years. The problem is to find out the total sum of savings after a certain number of years. A program can be written as follows.

```
[savings.py]
print '[Savings Calculator]'
rate = float(raw_input("Please enter the interest rate
    :"))
years = int(raw_input("Please enter the number of
    years:"))
init = float(raw_input("Please enter the initial sum
    of money: $"))
final = init * (1.0+rate)**years
print "After %d years, the savings will be $%.2f" % (
    years, final)
```

The program above, under the name "savings.py", consists of 6 lines of code. It can be typed in using any text editor, and saved into a specific folder in the file system. For the convenience of illustration, suppose that it is saved under the *Desktop* folder.

The IDLE editor can also be used to edit the program. To start editing a new program, go to the menu item [File→New Window], and a new text editor window will be popped up. Enter the code above into the editor window, and go to the menu item [File→Save...]. A dialog box will be opp up for entering the destination folder and file name. Choose *Desktop* as the destination folder and *savings.py* as the file name. A new Python code file *savings.py* will be created in the *Desktop* folder. It also appears on the Desktop of the graphic user interface of the computer.

In order to open and edit an existing file using IDLE, go to the menu item [File→Open...]. A dialog box will pop up for the selection of a specific file. In the case of this example, select the file *savings.py* from the *Desktop* folder. A new text editor window will pop up, which contains the current content of the file. Editing can be performed in the editor window.

In order to execute the current program in the editor, go to the menu item [Run→Run Module], or press the function key *F5*. The current window will be switched to the interactive IDLE window, and the following result can be shown according to the execution of the program.

```
>>> ============================== RESTART
    ==============================
[Savings Calculator]
Please enter the interest rate: 0.036
Please enter the number of years: 5
Please enter the initial sum of money: $10000
After 5\,years, the savings will be $11934.35
>>>
```

The program displays the title '[Savings Calculator]', and then asks the user to input the interest rate, the number of years after which the total savings should be calculated, and the initial sum of money. In this example, 0.036, 5 and 10000 are entered as the three values, respectively. The program then displays the result in the text console.

The Python program can also be executed directly using Python, under a text console. On a *Linux* or *Mac OS*, *Terminal* is the default text console.

```
Zhangs-MacBook-Pro:anywhere$ cd
Zhangs-MacBook-Pro:~ yue_zhang$ cd Desktop/
Zhangs-MacBook-Pro:Desktop yue_zhang$ python savings.
    py
[Savings Calculator]
Please enter the interest rate: 0.036
Please enter the number of years: 5
Please enter the initial sum of money: $10000
After 5\,years, the savings will be $11934.35
Zhangs-MacBook-Pro:Desktop yue_zhang$
```

On a *Windows OS*, the *Command Line* tool is the default text console.

```
C:\anywhere>cd %userprofile%
C:\Users\yuezhang>cd Desktop

C:\Users\yuezhang\Desktop>python saving.py
[Savings Calculator]
Please enter the interest rate: 0.036
Please enter the number of years: 5
Please enter the initial sum of money: $10000
After 5\,years, the savings will be $11934.35

C:\Users\yuezhang\Desktop>
```

In both examples above, the *cd* command introduced in Chap. 1 is used to change the current directory from an arbitrary location to the user's home directory, and then to the *Desktop* folder under this directory. Files in the folder are shown on the desktop of the operating system. **Direct execution** of a Python program is done by the command:

```
python <program file path>
```

Now back to the program itself. It consists of 6 statements. The first and last statements are two *print* statements, while the rest of the statements are assignment statements. With respect to functionality, the first four statements inquire the user for relevant inputs. The fifth statement calculates the answer based on user input, and the last statements shows the result to the user.

The Python program is executed line by line, from top to bottom. It can be treated as a list of statements, executed sequentially in a batch. The order is important, because the execution of one statement can depend on the result of another. For example, in the code above, *final* is calculated based on the user input of *init*. In subsequent chapters, more complicated dynamic execution orders will be introduced. But by default, Python always executes a program sequentially.

3.2.1 The Structure of Python Programs

The structure of a Python program can be summarized as Fig. 3.3, where a *program* consists of a set of *statements*. Each statement can contain one or more mathematical *expressions*, and each expression consists of *literals*, *identifiers* and *operators*. Python runs a program by executing its statements, each of which fulfills a specific functionality.

As mentioned earlier, all statements are executed in a fixed and mechanical way. Take the *assignment* statement introduced in the previous chapter for example. The right hand side of = must be an expression, and the left hand side must be an identifier (or a *tuple*, which will be introduced in Chap. 5). The statement is executed by evaluating the expression first, and then binding the resulting object with associating the left-hand-side identifier. Understanding the execution of each type of statement is crucial to the understanding and design of Python programs. A list of the statements that have been introduced up to this chapter is shown in Table 3.3.

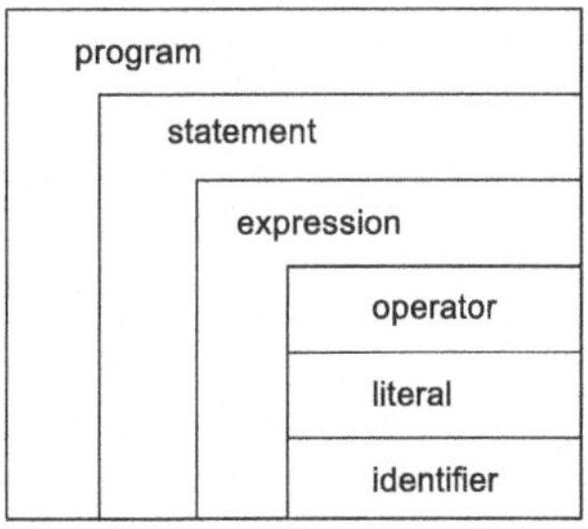

Fig. 3.3 The structure of a Python program

Table 3.1 A summary of types and literals

Type	Literal	Comment
int	1, 0, −1	Integer
long	1L	Implicit large integer
float	3.5, 5e−1	Floating point number
complex	1−3j	Complex numbers
str	'abc', "1 !"	For input output

Table 3.2 A summary of operators

Operator	Comment
+, −, *, /, //, %, **	Arithmetic operators
.	Load content from module
[]	getitem, getslice
()	Function call

Table 3.3 A summary of statements

Statement	Comment
assignment	*<identifier>*=*<expression>*, *<identifier>*+=*<expression>*,
import	Import module
print	Text output

Note that in the previous chapter, an expression is sometimes used as a command by itself. For example, by typing '3+5' in IDLE, one obtains the result '8'. However, as stated earlier, the usage above is possible only because IDLE displays the value of an expression by default. In a standalone Python program, this is not possible.

Each expression is evaluated into a single object during program execution. An expression is evaluated by applying operators to operands, which can be literals or identifiers. Operators are applied according to a fixed **precedence**, and brackets can be used to7 explicitly specify the order of evaluation. Table 3.2 give a summary of all the operators that have been introduced, which include arithmetic operators (+, -, * etc.), the dot (.) operator, the indexing operator ([]) and the function call operator (()). Their precedence is as follows. The dot, indexing and function call operators have higher precedence than the arithmetic operators, while ** has higher precedence than *, /, // and % in arithmetic operators. Operators with the same precedence are executed from left to right.

Both identifiers and literals represent Python objects in the underlying implementation. Each Python object is associated with a type. Two categories of types have been introduced in this book, including numbers (int, float, long, complex) and strings. While numbers are essential for mathematical computation, strings are useful for text input and output functionalities. A list of the types that have been introduced and their corresponding literals is shown in Table 3.1. More types will be introduced in subsequent chapters.

Comments. In addition to statements, a Python program can also consist of *comments*. Comments are not executed when a Python program is executed, but they are useful for making Python programs more readable. It is a good habit to write comments when programming, since proper comments can not only help other programmers understand the design of a program, but also remind a programmer of her own thinking when designing the program. Comments are particularly important in large software projects.

Python comments start with a # symbol. In a line of Python code, all the texts after the first # symbol are treated as comments, and ignored during execution.

```
>>> g=9.81 # the gravity constant
>>> y=g #+1
>>> print y #*2#*3
9.81
```

In the example above, the first line contains some comments that explain the purpose of the identifier g. The comments are useful for the understanding of subsequent uses of the identifier. The second line in the example above illustrates another use of comments, which is to temporarily delete a part of Python code. In this example, the value of y is 9.81, rather than 10.81, since the text '+1' is placed after a # symbol. In this case, the programmer might have started with the statement $y = g + 1$, but then decided to try $y = x$ instead. She, however, does not want to completely remove the '+1' part, since there is a chance that she wants to add it back after some testing. Therefore, she chooses to keep '+1' in the line but as a comment. In programming terminology, she *commented out* the part of code. The last line above contains two # symbols. It prints 9.81 on the console, but not 19.62 or 29.43, because all the text after the first # symbol are treated as comments.

Comments can also be written as a single line, with the # symbol at the beginning of the line. This form of comment is called **in-line comment**. In contrast, comments after a statement is called an **end-of-line comment**. In-line comments are typically used to explain the design of a set of multiple statements under the comment line.

Combined Statements. Multiple Python statements can be written in a single line, separated by semi-colons (;)

```
>>> a=1;b=2; print a #three statements combined
1
>>> b
2
```

Such combinations offer compactness in code, but can make programs more difficult to understand.

3.3 The Underlying Mechanism of Module Execution

Among the three types of statements in Table 3.3, the underlying mechanism of the assignment statement has been introduced in much detail. The *print* statement has been introduced in this chapter; it converts the value of an expression to string, and displays the string on the text console. The *import* statement, however, has been introduced only briefly. This section gives more details on the underlying mechanism of modules, which are useful for a better understanding of Python programs.

3.3.1 Module Objects

First, it has been introduced in the previous chapter that an imported module can be accessed by an identifier that bares the same name of the module (i.e. *math*). In addition, constants and functions defined in a module can be accessed by using the module name and the dot (.) operator (i.e. *math.e*). The former functionality is achieved by registering the name of the module in the binding table. To achieve the latter functionality, an imported module is associated with a binding table of its own, which contains the names of constants and functions defined in the module. For example, the memory structure after the following three lines of code are executed is shown in Fig. 3.4.

```
>>> x = 1
>>> y ='abc'
>>> import math
```

There are three identifiers in the main binding table: *x*, *y* and *math*. While *x* and *y* are put into the binding table by the corresponding assignment statements, *math* is inserted by the *import* statement. *x* and *y* are associated with the objects 1 and ‘abc’, respectively. *math*, in contrast, is associated with a binding table, which contains the constants and functions defined in the *math* module. In the figure, the name *sqrt* is associated with a function object, the structure of which will be introduced in Chap. 6.

The dot (.) operator syntactically connects a module name identifier to a second identifier (e.g. *math.e*). The *<module>.<identifier>* expression is evaluated to the value of *identifier*, looked up in the binding table of *module*. For example, when the expression *math.e* is evaluated, Python first looks for the name *math* in the main binding table. If it is found, Python follows the corresponding link to find the binding table of *math*, in which is searches for the identifier *e*. In each of the two look-ups, if the corresponding identifier is not found, an error will occur.

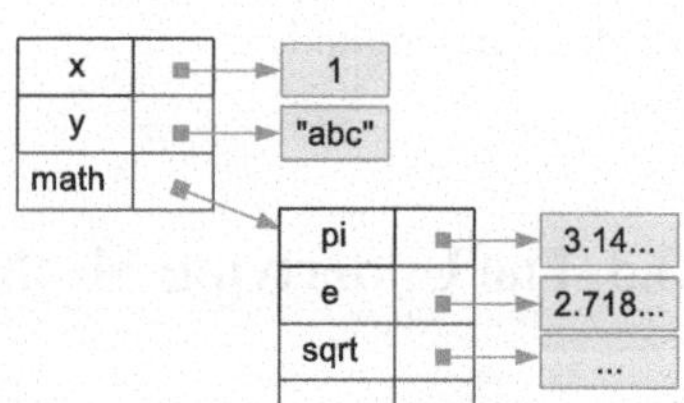

Fig. 3.4 Module name and binding table

```
>>> import math
>>> math.e
2.718281828459045
>>> math.e1
Traceback (most recent call last):
  File"<stdin>", line 1, in <module>
AttributeError: 'module' object has no attribute 'e1'
>>> math1.e
Traceback (most recent call last):
  File "<stdin>", line 1, in <module>
NameError: name 'math1' is not defined
```

In the example above, the second command (*math.e*) successfully returns the value of *math.e*, since *math* has been imported to the current binding table and *e* is a part of the *math* binding table. However, the third command (*math.e1*) raises an error because the name *e1* is not in the *math* binding table, and the fourth command (*math1.e*) raises an error because the name *math1* is not in the current binding table. In Python, the dot (.) operator indicates a binding table, specified by the left operand. An identifier is looked up in the specified binding table if it is the right operand of a dot operator (e.g. *math.e*), but in the current binding table otherwise (e.g. *e*). The main binding table of IDLE or a Python program is also called the **global binding table**.

3.3.2 Library Modules

The Python modules that have been introduced in this book are used as *libraries*, which are assemblies of code that provide specific functionalities for reuse. Most Python Library modules are nothing but normal Python programs, typically stored in a specific library path in the file system. The *random* module, for example, is a Python program that can be found on a *Linux* or *Mac OS* at:

```
/usr/lib/python2.7/random.py
```

and on a *Windows* system at:

```
C:\Python27\Lib\random.py
```

Most libraries provided by the Python distribution are located in the folders above. In addition to the Python distribution itself, libraries can also be downloaded from a third party. For example, the *SciPy* toolkit contains libraries for many scientific computation functionalities.[1] After installation, such libraries are typically located on a *Linux* or *Max OS* at:

```
/Library/Python/2.7/site-packages/
```

and on a *Windows* platform at:

```
C:\Python27\Lib\site-packages\
```

[1]http://www.scipy.org/.

Python automatically searches for the locations above when a library module is imported, where the folder names depend on the Python version. Note that the *math* module is an exception: it cannot be found under the library folder locations above. This is because *math* is a very commonly used module, and is built in the main Python program itself. Other commonly used modules, including *cmath*, are also built-in.

Making Use of Libraries. Before writing one's own code for a specific functionality, it is wise to check whether there is an existing library module that can be used directly. Python provides many powerful functionalities through modules distributed along with the Python program itself, and the descriptions of all the modules can be found in the Python documentation. To look for third-party modules with given functionalities, online search engines are a useful resource. For example, searching for 'Python scientific computation' can lead to the website of *SciPy*.

3.3.3 The Mechanism of Module Importation

The *import* statement creates an identifier that has the same name as the module. For example, by using *'import random'*, *'random.py'* is imported into the memory, and associated with the name '*random*' in the global binding table. To avoid name clashes and for the convenience of reference, a module can also be given a different name when imported into the binding table. This can be achieved by using a variant of the *import* statement, with the syntax '*import x as y*', where *x* is the name of the module in the file system and *y* is the desired name in the binding table.

```
>>> import math as m
>>> m.e
2.718281828459045
>>> m.sqrt(25.0)
>>> 5.0
>>> math.e
Traceback (most recent call last):
  File "<stdin>", line 1, in <module>
NameError: name 'math' is not defined
```

In the example above, the *math* module is loaded into the binding table and associated with the identifier *m*. As a result, by using "*m.* ", the corresponding binding table of the *math* module can be accessed. However, the identifier "*math*" does not exist in the global binding table this time.

Since both the assignment statement and the *import* statement changes the binding table, they can override the value of an identifier. For example,

```
>>> x=1
>>> import math as x
>>> x+1 # not a number
Traceback (most recent call last):
  File"<stdin>", line 1, in <module>
TypeError: unsupported operand type(s) for +: 'module'
    and 'int'
>>> x.pi # but a module
```

```
3.141592653589793
```

In the example above, *x* is initially bound to a number, but then rebound to a module. The original binding is *overridden*. An assignment statement can also override an import statement. For example,

```
>>> import math
>>> math=1
>>> math+1 # number
2
>>> math.e # error
Traceback (most recent call last):
  File "<stdin>", line 1, in <module>
AttributeError: 'int' object has no attribute 'e'
```

As shown earlier, modules such as *random.py* are Python programs themselves. In fact, *any Python program* can be loaded as a module. To verify this, the program *saving.py* can be used as an example. Suppose that another source file, *textimport.py*, is created under the *Desktop* folder. It contains the following two lines of code

```
[testimport.py]
import savings
x=1
print savings.rate, "is entered as the interest rate"
```

Executing the program above yields the following result.

```
Zhangs-MacBook-Pro:~ yue_zhang$ cd Desktop/
Zhangs-MacBook-Pro:Desktop yue_zhang$ python
   testimport.py
[Savings Calculator]
Please enter the interest rate: 0.036
Please enter the number of years: 5
Please enter the initial sum of money: $10000
After 5 years, the savings will be $11934.35
0.036 is entered as the interest rate
Zhangs-MacBook-Pro:Desktop yue_zhang$
```

One immediate thing to notice is that the program asks the user for the interest rate, number of years and initial deposit again, just as if the *savings.py* program were executed alone. Then it displays the message "0.036 is entered as the interest rate". Here, *savings.rate* is used to access the *rate* value defined in savings.py, in the same way as the use of *math.e* to access the value of *e* defined in *math*.

The underlying mechanism of module importation is reflected in this example. First, the importing module hands control over to the imported module, which is executed, with its own binding table being the global binding table. Second, the importing module resumes control, with the global binding table switched back to the binding table of the importing module. A new entry added to the binding table of the importing module, associating the name of the imported module with the binding table of the imported module. This process can be illustrated in Fig. 3.5. The example also shows **a second use of the print statement**, where a comma-separated list of expressions is put after the keyword. In this case, each expression will be printed, with a space separating the outputs.

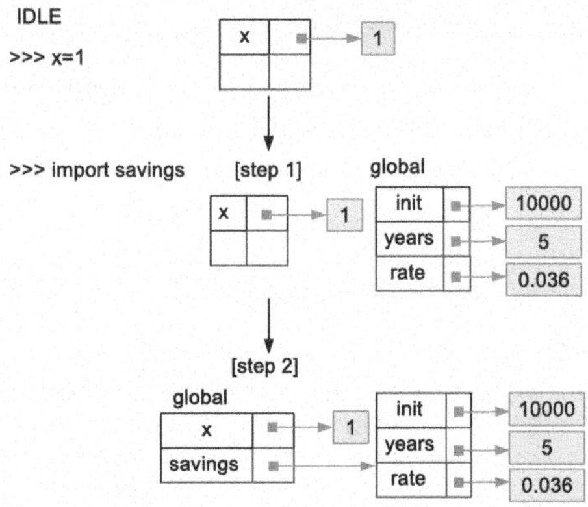

Fig. 3.5 Example memory structure for module import

3.3.4 Duplicated Imports

If the same module is imported twice, the content of the module will not be executed by the second *import* statement. For example, the following change to the *testimport.py* code will not lead to any change in the runtime behavior.

```
[testimport.py]
x=1
import savings
import savings
print savings.rate,"is entered as the interest rate"
```

The first two lines of the program remain the same as before, while the third line is added for a second import of the same module. When this program is executed, the user will be asked for input only once, with the same value of *savings.rate* being shown when the fourth line of code is executed. Executing *testimport.py* from IDLE can yield the following output.

```
>>> ================================ RESTART
    ================================
[Savings Calculator]
Please enter the interest rate: 0.036
Please enter the number of years: 5
Please enter the initial sum of money: $10000
After 5 years, the savings will be $11934.35
0.036 is entered as the interest rate
>>>
```

It can be seen in the example above that, no operation is taken when the second *import* statement is executed. Note that for the example to run as expected, *savings.py* must be in a folder that is accessible by IDLE. For example, if the *Desktop* folder has been added to the *PYTHON_PATH* environment variable as illustrated in Chap. 1, all the files in the folder are accessible. Alternatively, if IDLE is started at the *Desktop* folder, all the files in the folder are also accessible. The following example shows how this can be done on a *Linux* or *Max OS*.

```
Zhangs-MacBook-Pro:anywhere$ cd
Zhangs-MacBook-Pro:~ yue_zhang$ cd Desktop/
Zhangs-MacBook-Pro:Desktop yue_zhang$ idle
```

The following example shows the case on a *Windows OS*.

```
c:\anywhere> cd %userprofile%
c:\Users\yue_zhang> cd Desktop/
c:\Users\yue_zhang\Desktop> idle
```

Two imports of the same module behave the same as in the previous example even when they are executed from different modules—execution of the second *import* statement does not result in the module being executed again. For example, suppose that the following two files are in the *Desktop* folder.

```
[testimport1.py]
import savings
x=1

[testimport2.py]
import savings
import testimport1
y=1
```

When *testimport2.py* is executed as the main program, the dynamic execution sequence is:

```
import savings
[execute savings.py]
   ...
[savings.py finished]
import testimport
[execute testimport.py]
   import savings
   [execute savings.py]
      ...
   [savings.py finished]
   x=1
   [testimport.py finished]
y=1
```

savings.py is executed only once, when the first line of *testimport2.py* is executed. When the second line of *testimport2.py* is executed, *testimport1.py* is executed. But when the first line of *testimport1.py* is executed, *savings.py* is not executed a second time. The same module object is simply associated with a new entry in the binding table of *testimport1*. After the execution of $y = 1$, the memory structure is shown in Fig. 3.6. Note that the same module object is associated with the name *savings* in two different binding tables.

Reloading a module. In case the imported module is modified between the execution of two *import* statements, the second *import* will not reflect the modification. Python has a special function, *reload*, which reloads the context of an imported module by executing it once more. For example, suppose that *savings.py* is changed between an *import* statement and a *reload* function call as follows:

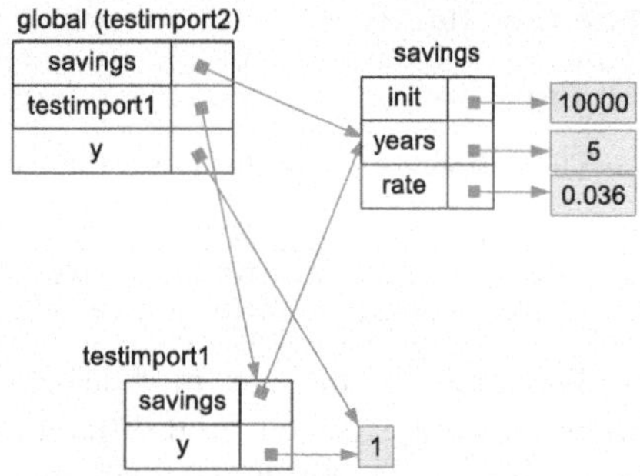

Fig. 3.6 Example memory structure for duplicated import

```
[savings.py initially]
print '[Savings Calculator]'
rate = float(raw_input("Please enter the interest rate:"))
years = int(raw_input("Please enter the number of years:"))
init = float(raw_input("Please enter the inital sum of money: $"))
final = init * (1.0+rate)**years
print "After %d years, the savings will be $%.2f" % (years, final)

[savings.py after import but before reload]
print 'The new version'
init=0
```

Execution of the *import* and *reload* operations yields the following result:

```
>>> import savings
[Savings Calculator]
Please enter the interest rate: 0.036
Please enter the number of years: 5
Please enter the initial sum of money: $10000
After 5 years, the savings will be $11934.35
0.036 is entered as the interest rate
>>> import savings   # nothing happens
>>> reload (savings)
The new version
>>> print savings.init
0
```

The *reload* function changes the context of the binding table of *savings*, where the value of *init* becomes 0. As a result, 0 is printed by the last command in IDLE.

3.3.5 Importing Specific Identifiers

Sometimes only a small number of functions or variables needs to be imported from a module. In such cases, the use of a module name in reference to the imported functions or variables may be unnecessary. For example, suppose that only *pi* and *sin* are needed by a program. In this case, importing the two identifiers directly into the global binding table can be more convenient than importing the *math* module, because in this way *pi* and *sin* can be referred to directly, instead of using *math.pi* and *math.sin*.

Python provides a variation of the *import* statement for such imports. The syntax is:

```
from <module name> import <identifiers>
```

where *<identifiers>* is a comma-separated list of specific identifiers, or a wildcard * that represents all the identifiers in the module.

In the following example, the variable *pi* and the function *sin* are imported from the *math* module directly.

```
>>> from math import pi, sin
>>> sin(pi)
1.2246467991473532e-16
>>> math.pi
Traceback (most recent call last):
  File "<stdin>", line 1, in <module>
NameError: name 'math' is not defined
```

In the example above, the module name *math* cannot be found in the global binding table, but the identifiers *pi* and *sin* are present. The mechanism for the *from ... import ...* statement is illustrated in Fig. 3.7. It consists of two steps: (1) execute the module to be imported; (2) add the specified identifiers to the global binding tables. The process is the same as the *import* ... statement in the first step, but differs in the second step. As shown in Fig. 3.7, the resulting global binding table contains the identifiers *pi* and *sin*, but not the identifier *math*. The binding table of the *math* module is not referred to by any identifier in this case, and will be removed by the garbage collector, together with the objects in the module other than *pi* and *sin*.

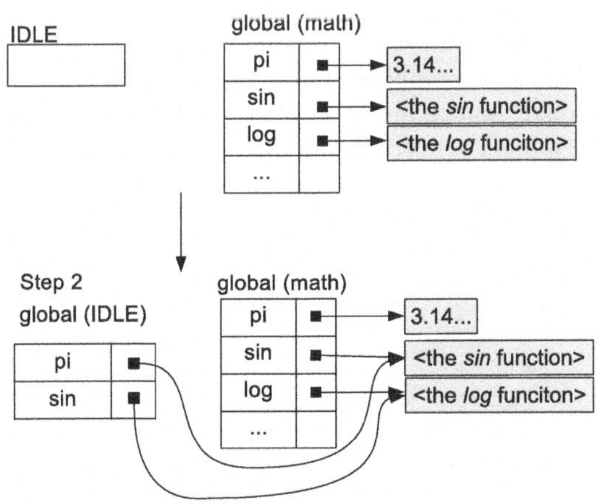

Fig. 3.7 Memory structure of *from ... import ...*

The following example shows the importing of all identifiers from the *math* module.

```
>>> from math import *
>>> sin(pi/2)
1.0
>>> log(e)
1.0
>>> factorial(10)
3628800
>>> pow(2, 8)
256.0
```

dir **and the** ***__builtins__*** **binding table**

Python allows the listing of all entries in the global binding table by using the *dir* function.

```
>>>dir()
['__builtins__', '__doc__', '__name__', '__package__']
>>> x=1
>>> y=2
>>> dir()
['__builtins__', '__doc__', '__name__', '__package__',
    'x', 'y']
```

In the example above, the first *dir* function call gives four entries, *__builtins__*, *__doc__*, *__name__* and *__package__*, which are the default entries when IDLE (or a Python program) starts. Details about these entries will be introduced later in Chap. 11. When the two assignment statements $x = 1$ and $y = 2$ have been executed, two more entries, x and y are added to the global binding table, as shown by the second call to the *dir* function.

The *dir* function can also take an input argument, which is an object that has a binding table, and lists out all the entries in the binding table. As introduced earlier, module objects are associated with binding tables. As a result, one can list out all the identifiers defined in a module.

```
>>>import math
>>>dir(math)
['__doc__', '__name__', '__package__', 'acos', 'acosh'
    , 'asin', 'asinh', 'atan', 'atan2', 'atanh', 'ceil'
    , 'copysign', 'cos', 'cosh', 'degrees', 'e', 'erf',
    'erfc', 'exp', 'expm1', 'fabs', 'factorial', '
    floor', 'fmod', 'frexp', 'fsum', 'gamma', 'hypot',
    'isinf', 'isnan', 'ldexp', 'lgamma', 'log', 'log10'
    , 'log1p', 'modf', 'pi', 'pow', 'radians', 'sin', '
    sinh', 'sqrt', 'tan', 'tanh', 'trunc']
```

In the list above, the first three entries are special and exist for all modules; more about them will be discussed in Chap. 11.

Note that the input argument to the *dir* function call must be available in the current global binding table. Otherwise, an error will be given.

```
>>>dir(random) # random has not been imported
```

```
Traceback (most recent call last):
  File "<pyshell#6>", line 1, in <module>
    dir(random)
NameError: name 'random' is not defined
```

In addition to modules, other objects can also be associated with binding tables, and they will be introduced in Chap. 10.

It can be noticed that the global binding table does not contain built-in functions such as *id*, *int*, *len* and *dir*. The reason that they can be referred to without causing an error is that Python organizes all built-in functions in a special module, *__builtins__*, which is implicit in the system. When an identifier cannot be found in the global binding table, the *__builtins__* binding table will be searched. In fact, the *__builtins__* in the global binding table by default, as shown earlier.

```
>>> dir(__builtins__)
['ArithmeticError', 'AssertionError', 'AttributeError'
    , 'BaseException', 'BufferError', 'BytesWarning', '
    DeprecationWarning', 'EOFError', 'Ellipsis', '
    EnvironmentError', 'Exception', 'False', '
    FloatingPointError', 'FutureWarning', '
    GeneratorExit', 'IOError', 'ImportError', '
    ImportWarning', 'IndentationError', 'IndexError', '
    KeyError', 'KeyboardInterrupt', 'LookupError', '
    MemoryError', 'NameError', 'None', 'NotImplemented'
    , 'NotImplementedError', 'OSError', 'OverflowError'
    , 'PendingDeprecationWarning', 'ReferenceError', '
    RuntimeError', 'RuntimeWarning', 'StandardError', '
    StopIteration', 'SyntaxError', 'SyntaxWarning', '
    SystemError', 'SystemExit', 'TabError', 'True', '
    TypeError', 'UnboundLocalError', '
    UnicodeDecodeError', 'UnicodeEncodeError', '
    UnicodeError', 'UnicodeTranslateError', '
    UnicodeWarning', 'UserWarning', 'ValueError', '
    Warning', 'WindowsError', 'ZeroDivisionError', '_',
     '__debug__', '__doc__', '__import__', '__name__',
    '__package__', 'abs', 'all', 'any', 'apply', '
    basestring', 'bin', 'bool', 'buffer', 'bytearray',
    'bytes', 'callable', 'chr', 'classmethod', 'cmp', '
    coerce', 'compile', 'complex', 'copyright', '
    credits', 'delattr', 'dict', 'dir', 'divmod', '
    enumerate', 'eval', 'execfile', 'exit', 'file', '
    filter', 'float', 'format', 'frozenset', 'getattr',
     'globals', 'hasattr', 'hash', 'help', 'hex', 'id',
     'input', 'int', 'intern', 'isinstance', '
    issubclass', 'iter', 'len', 'license', 'list', '
    locals', 'long', 'map', 'max', 'memoryview', 'min',
     'next', 'object', 'oct', 'open', 'ord', 'pow', '
    print', 'property', 'quit', 'range', 'raw_input', '
    reduce', 'reload', 'repr', 'reversed', 'round', '
    set', 'setattr', 'slice', 'sorted', 'staticmethod',
     'str', 'sum', 'super', 'tuple', 'type', 'unichr',
    'unicode', 'vars', 'xrange', 'zip']
```

There are many built-in functions, such as *round* and *len*; many more will be introduced in this book. By convention, Python names implicit or special objects by enclosing their name in double underscores (__).

Finally, note that the *__builtins__* module is searched only when a reference to the *global* binding table is not found. Identifiers in a *specified* binding table will not be looked for in the *__builtins__* binding table.

```
>>> import math
>>> math.dir()


Traceback (most recent call last):
  File "<pyshell#9>", line 1, in <module>
    math.dir()
AttributeError: 'module' object has no attribute 'dir'
```

In the example above, because of the use of *math.*, Python searches for the identifier *dir* only in the *math* module. When it is not found, Python raises an error without searching the *__builtins__* module for the name *dir*.

Exercises

1. Does the following expression result in the string object '*abc*'? Why?

   ```
   str(abc)
   ```

2. Write a Python program that

   (a) asks the user for a floating point number x_0, and prints out the value of $f(x) = x^3 - 3x + 1$ at x_0.
   (b) asks the user for her name, and then prints "*Hello, <name>!* ", when *<name>* is the name of the user.
   (c) asks the user for an integer, and then displays the number of digits in the integer, and the logarithm of the integer.
   (d) asks the user for an integer, and then displays the number of digits in the integer to the power of 3.
   (e) asks the user for a string *s*, and displays the following string: '<s>***<s> ***<s>', where <s> is the string given by the user.
   (f) asks the user for a string, and displays a string that repeats the input string *n* times, where *n* is the length of the input string.

3. Write a program that asks the user for her name, and a short message below 50 characters, displaying the following output

```
****************************************
*                                      *
*             Hello, <name>            *
*                                      *
*            Your message is:          *
*                                      *
* <The content of the message>         *
* <maybe continued message>            *
****************************************
```

In the output above, each line contains 40 characters. *<name>* is the name of the user. The string 'Hello, <name>' must be centered in the third line, regardless of the length of *<name>*. The content of the message starts from the third character of the seventh line. If it is more than 36 characters, the text should be wrapped into the eighth line. Otherwise the eighth line is empty.

4. Write a program that asks for user input and performs
   (a) Celsius to Fahrenheit conversion.
   (b) Fahrenheit to Celsius conversion.
   (c) Meter to foot conversion.
   (d) Foot to meter conversion.
   (e) Acre to square meter conversion.
   (f) Square meter to acre conversion.
   (g) Pound to kilogram conversion.
   (h) Kilogram to pound conversion.
5. Make the answers to question 4 in the previous chapter full programs, which allow arbitrary user input, giving the corresponding results.
6. Draw the binding tables after the following code is executed. How many times are the *math* module executed?

```
[lib.py]
import math
x=math.pi/2

[main.py]
import lib
import math
y=math.cos(lib.x)
print y
```

Chapter 4
Branching and Looping

Many a time, the computer needs to take different actions according to user input. For example, when the user closes a text editor window, the operating system prompts out a dialog box asking the user whether to save her work before quitting. According to the user choice, the program may save the work before exiting, or exit without saving the work. Unlike the programs in the precious chapter, the **dynamic**, or **runtime** execution sequence of the text editor program can be different from the **static** lines of code written in the program. In the example above, the code for saving the user's work may or may not be executed at runtime. This is called **conditional** execution, or **branching**.

Another type of dynamic behavior is **looping**, where the computer repeats the same process multiple times. For example, a music player can repeatedly play the same song multiple times. In a dynamic execution, the same piece of static code (i.e. play music) is executed multiple times. In addition, the number of times the same code is repeated can also depend on user input. The playing of a playlist is also typically implemented using loops, by repeating the *same* music playing code on *different* songs.

The execution of a program can be treated as a continuous flow of statement execution. Branching and looping can be treated as two different ways of **control flow**, which refers to how a sequence of static code is put into a dynamic sequence for execution at runtime. In contrast to branching and looping, the execution sequence introduced in the previous chapters, where a list of statements are executed one by one, is also referred to as the *sequential* control flow. In a sequential control flow, the order in which a set of statements is executed is exactly the same as the order in which they are written.

Figure 4.1 shows a comparison between sequential, branching and looping execution sequences. This type of diagram is called **control flow diagram**, or **flow charts**. Flow charts consist of arrows and nodes. While arrows indicate the execution sequence, nodes can be either statements or junctions. A junction is a binary branching node, depending on a condition, it determines while branch to follow in dynamic execution. It turns out that a combination of the three basic types of control

Y. Zhang, *An Introduction to Python and Computer Programming*,
Lecture Notes in Electrical Engineering 353, DOI 10.1007/978-981-287-609-6_4

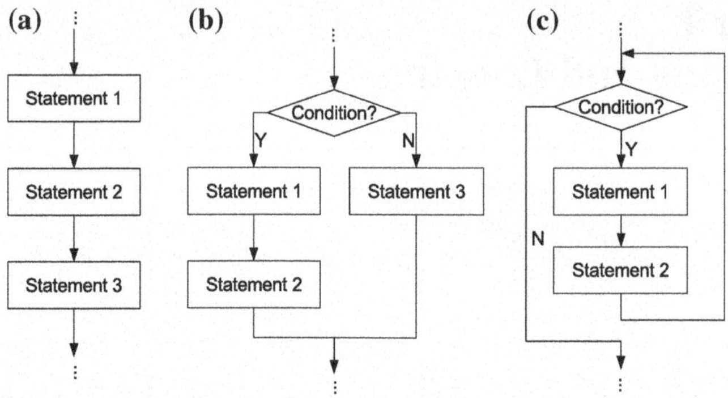

Fig. 4.1 The three basic types of control flow. **a** Sequential. **b** Branching. **c** Looping

flow in Fig. 4.1 is sufficient to solve any *computable* problem—any problem that can be solved by a computer.

The understanding of branching and looping is a milestone in learning programming. It is crucial to be able to tell the difference between dynamic execution and static code when these two types of control flow are involved. For both branching and looping in Fig. 4.1, there is a control junction that chooses between two branches of execution according to a binary option. The option can be expressed using a **Boolean** type expression in Python.

4.1 The Boolean Type

The Boolean type is designed to represent logical values; it is frequently used for control flow. Different from integers, floating point numbers and strings, for which there are an infinite number of possible literals, there are only two **Boolean literals**: *True* and *False*.

```
>>> a=True
>>> type(a)
<type 'bool'>
>>> b=False
>>> type(b)
<type 'bool'>
```

Boolean operators. There are three logical operators for Boolean objects, which include the binary operators *and* and *or*, and the unary operator *not*. The effects of the three operators are shown in Fig. 4.2. *a and b* is *True* only when both *a* and *b* are *True*, and *False* otherwise. *a or b* is *False* only when both *a* and *b* are *False*, and *True* otherwise. *not a* is *True* when *a* is *False*, and *False* when *a* is True.

(a)

op1 \ op2	True	False
True	True	False
False	False	False

(b)

op1 \ op2	True	False
True	True	True
False	True	False

(c)

op	
True	False
Flase	True

Fig. 4.2 Boolean operators. **a** and. **b** or. **c** not

```
>>> a=True
>>> b=False
>>> a and b
False
>>> a or b
True
>>> not a
False
>>> not b
True
```

Similar to arithmetic expressions, Boolean expressions can be composite, consisting of multiple Boolean objects and operators. The priority of the *not* operator is higher than that of *and*, which is in turn higher than that of *or*. Similar to other expressions, brackets can be used to explicitly specify the order of evaluation in Boolean expressions.

```
>>> a=True
>>> b=True
>>> c=False
>>> a and not b              # 'not b' first
False
>>> c and b or a             # 'c and b' first
True
>>> c and (b or a)           # explicit
False
>>> a and b or b and c       # 'a and b'→'b and c'→'or'
   last
True
```

In the last example above, *a and b* is evaluated first, before *b and c* is evaluated. The resulting values *True* and *False* are then combined by the *or* operator, resulting in the value *True*. If *and* and *or* had the same precedence, the expression would have been evaluated from left to right, resulting in the value *False*. With two literals and three operators, Boolean algebra is relatively simple and straightforward.

Operators resulting in Boolean values. There is a set of operators that take non-Boolean operands, yet result in Boolean values. One example of such operators is **numerical comparison**.

```
>>> 1==2                    # equals
False
>>> 1!=2                    # not equal
True
>>> 3>5                     # greater than
False
>>> 3<5                     # less than
True
>>> 3<=5                    # less than or equal to
True
```

The == operator (i.e. =; equal) returns *True* if the two operands are equal, and *False* otherwise; the != operator (i.e. ! =; unequal) returns *True* if the two operands are not equal, and *False* otherwise; the > operator (i.e. greater than) returns *True* if the first operand is greater than the second operand, and *False* otherwise; the < operator (i.e. less than) returns *True* if the first operand is less than the second operator, and *False* otherwise; the >= operator (i.e. >=; greater than or equal to) returns *True* if the first operand is greater than or equal to the second operand, and *False* otherwise; the <= operator (i.e. <=; less than or equal to) returns *True* if the first operand is less than or equal to the second operand, and *False* otherwise. In addition to the operators above, more operators take non-Boolean operands and result in Boolean values; they will be introduced later in this book. Intuitively, the two special floating point numbers, *float('inf')* and *float('-inf')*, which represent ∞ and $-\infty$, respectively, conform to the rules of numerical comparison also.

```
>>> a=float('inf')
>>> b=float('-inf')
>>> a>10e6
True
>>> a<10
False
>>> b<-10E6
True
```

Multiple Boolean expressions can be combined by Boolean operators, resulting in composite Boolean expressions, which are commonly used in Python programs for control flow.

```
>>> a=3
>>> b=5
>>> c=2
>>> a>b and b>c
False
>>> a>b or b>c
True
>>> a>b and b>c or b>c and c<a
True
```

Operators that result in Boolean values (e.g. >, <) have higher precedence compared to Boolean operators (e.g. *and*, *or*). As a result, the three expressions in the example above are reduced to composite Boolean expressions after the > and < operators are evaluated. In the first two examples above, the expressions $a > b$ and $b > c$ are evaluated first, before the Boolean operators are evaluated. In the last example, the four expressions $a > b$, $b > c$, $b > c$ and $c > a$ are evaluated first, before the two *and* operators are evaluated. The *or* operator is evaluated last due to the fact that *and* has higher priority than *or*.

The operators ==, !=, <, >, <= and >= can also be cascaded, with the interpretation of logical conjunction. For example, $1 < 2 < 3$ is interpreted as $1 < 2$ *and* $2 < 3$, and its value is *True*. Similarly, the value of $1 <= 2 < 5! = 4$ is *True*. However, the value of $1 < 3 < 5 == 4$ is *False* because the value of $5 == 4$ is *False*.

Another set of operators that result in Boolean values include *in* and *not in*, which return whether a string is a sub string of another.

```
>>> s1 = 'bc'
>>> s2 = 'abcde'
>>> s3 = 'b'
>>> s1 in s2
True
>>> s3 not in s2
False
>>> s3 in s1
True
>>> s1 in s3
False
```

Yet another set of operators that result in Boolean values includes *is* and *is not*, which return whether the identifier of the object are the same.

```
>>> a = 123          # small number cached
>>> b = 123
>>> id(a)
19960736
>>> id(b)
19960736
>>> a is b
True
>>> a is not b
False
>>> a = 12345        # not cached
>>> b = 12345
>>> id(a)
20241576
>>> id(b)
20241408
>>> a is b
False
```

Type conversion from and to the Boolean type can be applied to numbers, strings and most other types to be introduced in this book.

```
>>> bool(0)        # zero
False
>>> bool(3)        # non-zero
True
>>> bool(0.0)      # zero
False
>>> bool(0.1)      # non-zero
```

```
True
>>> bool(1.0)    # non-zero
True
>>> bool(100.0) # non-zero
True
>>> bool('')     # empty
False
>>> bool('abc') # non-empty
True
>>> bool('This is a string that contains multiple
    characters.')
True
>>> int(True)
1
>>> int(False)
0
>>> float(True)
1.0
>>> float(False)
0.0
>>> str(True)
'True'
>>> str(False)
'False'
>>> print False
False
>>> print True
True
```

While the *bool* function converts objects of other types into Boolean values, Boolean values can be converted into other types using respective functions. In general, a number is converted to *False* if its value is 0, and *True* otherwise. A string is converted to *False* if it is an empty string, and *True* otherwise. The numeric value of *False* is 0, and that of *True* is 1. The string value of *False* and *True* are 'False' and 'True', respectively.

The *and*, *or* and *not* operators require Boolean operands. If non-Boolean operands are found, they are converted to Boolean values implicitly. For example, (1 *and* 'a') is *True* because both 1 and 'a' are converted to *True*. However, 0 *and* 3 is *False* because 0 is converted to *False*.

4.2 Branching Using the *if* Statement

The *if* statement supports conditional or branching control flow. The simplest form of the *if* statement is shown in Fig. 4.3. It is a **compound statement** that consists of multiple lines. The first line contains a Boolean condition, the runtime value of which decides whether the rest of the compound statement is executed.

In particular, the first line starts with the *if* keyword, followed by a Boolean expression, and then a colon (:). Following the *if* line, a *block* of *indented statements* are written from the second line onwards. Here **indentation** refers to the insertion

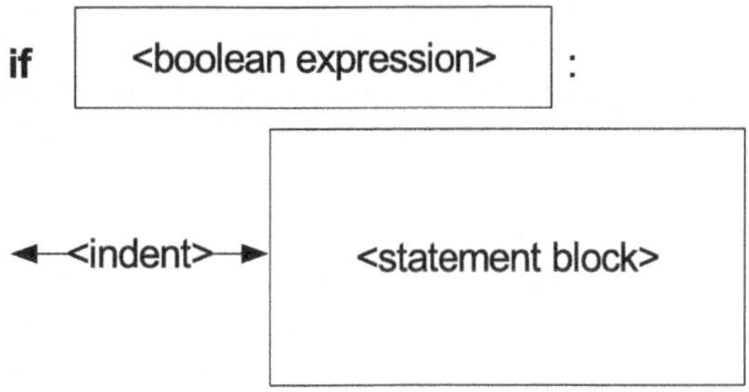

Fig. 4.3 The simplest form of the *if* statement

of whitespace characters, including spaces ('') and tabs (' ') before a statement. Common indentations include 2–8 spaces and 1–2 tabs, while spaces are preferred because different text editors can interpret a tab character as different numbers of spaces. A **statement block** contains one or multiple statements with equal identation.

Conditional execution. At runtime, the value of the Boolean expression in the *if* line determines whether the statement block is executed. For example,

```
>>> if True:
...     a=1
...     b=2
...     print a+b
...
3
>>> if False:
...     a=1
...     b=2
...     print a+b
...
>>>
```

There are two *if* statements in the example above. In the first *if* statement, the Boolean expression is a constant value *True*. When it is executed, the statement block under the *if* line is executed. In the second *if* statement, the Boolean expression is a constant *False*. When this *if* statement is executed, the statement block under the *if* line is not executed, and nothing is printed.

Below are three examples where the Boolean condition expressions are more complex than constant values.

```
>>> a=1
>>> b=2
>>> c=3
>>> if a>b:
...     print 'yes'
...     print 'a>b'
...
>>> if b<c:
...     print 'yes'
...     print 'b<c'
...
yes
b<c
```

```
>>> if a>b or b<c:
...     print 'yes'
...
yes
>>>
```

In the first example above, the Boolean expression consists of a > operator, which takes two integer operands and returns a Boolean value. According to the values of *a* and *b*, the value of the Boolean expression is *False*, and hence the two statements under the *if* line are not executed. In the second example, the same type of Boolean expression is used, but the runtime value is *True*, leading to the execution of the two indented statements. The third example uses a compound Boolean expression, which is evaluated to *True*, leading to the execution of the statement block.

If a non-Boolean expression is put in the place of the Boolean expression in the *if* statement, it is automatically and implicitly converted into a Boolean value.

```
>>> a=1
>>> b=-2.0
>>> c=0
>>> d='abc'
>>> e=''
>>> if a:              #non-zero
...     print a
...
1
>>> if b:              #non-zero
...     print b
...
-2.0
>>> if c:              #zero
...     print c
...
>>> if d:              # non-empty
...     print d
...
abc
>>> if e:              # empty
...     print e
...
>>>
```

In the lines of code above, five values of three different types are put in the place of the Boolean expression. They are converted into the Boolean type, respectively. Again the outputs illustrate the mechanisms of Boolean conversion and conditional execution.

In a dynamic execution of the *if* statement, the value of the Boolean expression can be determined by user input.

```
[zero.py]
a=int(raw_input('Give me a number: '))
if a==0:
   print 'The number is zero'
print 'Bye'
```

The program asks the user for a number *a*, and prints a line if it is zero. This is achieved by using an *if* statement with the Boolean condition $a == 0$. After the compound *if* statement, there is a third statement in the program, which simply prints the string 'Bye'. Python uses indentation to determine whether a statement is *inside* a compound statement or *outside* it. In this example, the last line is not indented, and therefore is not a part of the compound *if* statement.

An example of executing the program is shown below.

```
Zhangs-MacBook-Pro:code yue_zhang$ python zero.py
Give me a number: 0
The number is zero
Bye
Zhangs-MacBook-Pro:code yue_zhang$ python zero.py
Give me a number: 1
Bye
```

Here when the user inputs a zero, the *print* statement inside the *if* statement is executed, and two lines of text are printed. When the user inputs a non-zero number, the *print* statement under the *if* statement is not executed. The flow chat of the program is shown in Fig. 4.4a.

Two-way branching. *zero.py* gives a text response when the user input is zero, but no feedback if the user input is non-zero. In order to be more informative and also show some text response when the user input is not zero, an alternative form of the *if* statement can be used.

```
[zero_else.py]
a=int(raw_input('Give me a number: '))
if a==0:
   print 'The number is zero'
else:
   print 'The number is not zero'
print 'Bye'
```

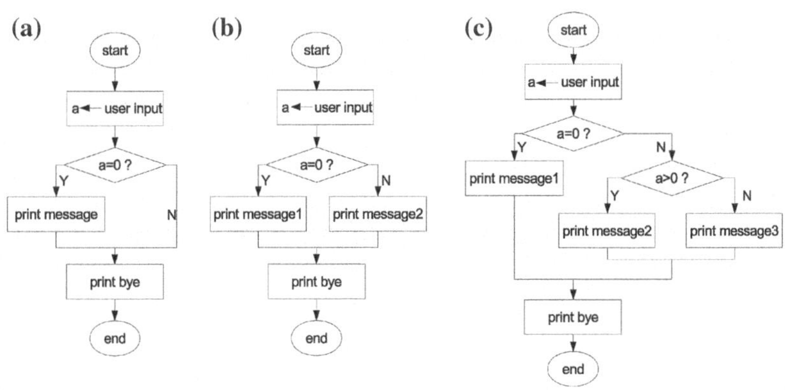

Fig. 4.4 Branching flow charts. **a** zero.py (if). **b** zero-else.py (if ... else ...). **c** zero-elif.py (if ... elif ... else ...)

zero_else.py is a modified version of *zero.py*, with the *if* statement being extended by adding an *else* component, which consists of two lines. The first line is not indented; it starts with the *else* keyword, followed by a colon (:). The second line is a *print* statement, which by itself forms an indented statement block.

The *if ... else ...* structure is a single compound statement, the execution of which is controlled by the Boolean expression in the *if* line. If the value of the Boolean expression is *True*, the indented statement block under the *if* line is executed, but the indented statement block under the *else* line is not. If the value of the Boolean expression is *False*, the reverse happens, with the indented statement block under the *else* line being executed, but that under the *if* line being *not* executed. At each execution of the compound statement, only one of the two indented statement blocks is executed.

An example of executing *zero_else.py* is shown below:

```
Zhangs-MacBook-Pro:code yue_zhang$ python zero_else.py
Give me a number: 0
The number is zero
Bye
Zhangs-MacBook-Pro:code yue_zhang$ python zero_else.py
Give me a number: 1
The number is not zero
Bye
Zhangs-MacBook-Pro:code yue_zhang$ python zero_else.py
Give me a number: 3
The number is not zero
Bye
```

As can be seen from the example above, the program gives a corresponding response regardless whether the user input is zero or non-zero. The dynamic control flow is illustrated in Fig. 4.4b.

Multi-way branching. *zero_else.py* can be further extended to give different responses when the user input is positive, zero or negative, respectively. This is a three-way branching condition. However, one Boolean value can express a choice between only two options. A common solution to making a choice from more than two options is to break a multi-way condition into the concatenation of multiple two-way conditions. In this case, two Boolean conditions can be used, with the first one determining whether the user input is zero or non-zero, and the second one determining whether the user input is positive or not, if it is non-zero. The dynamic process can be illustrated by Fig. 4.4c.

Python offers an alternative form of the *if* statement that allows such multi-way branching. Making use of this variation, *zero_else.py* can be further extended into *zero_elif.py*.

```
a=int(raw_input('Give me a number: '))
if a==0:
   print 'The number is zero'
elif a>0:
   print 'The number is positive'
else:
   print 'The number is negative'
print 'Bye'
```

zero_elif.py is different from *zero_else.py* in that the second statement, which is a compound *if ... else ...* statement in *zero_else.py*, is extended into an *if ... elif ... else ...* compound in *zero_elif.py*. The additional *elif* line corresponds to the additional condition in Fig. 4.4c. It is an unindented line, starting with the keyword *elif*, followed by a Boolean expression, and finishing with a colon (:). Similar to the *if* and *else* cases, an indented statement block is placed under the *elif* line. The dynamic execution of the compound *if ... elif ... else ...* statement can be illustrated with Fig. 4.4c, where the three branches from left to right correspond to the statement blocks under the *if*, *elif* and *else* lines, respectively. At each execution, only one among the three branches is executed. An example of executing *zero_elif.py* is shown below.

```
Zhangs-MacBook-Pro:code yue_zhang$ python zero_elif.py
Give me a number: 1
The number is positive
Bye
Zhangs-MacBook-Pro:code yue_zhang$ python zero_elif.py
Give me a number: 0
The number is zero
Bye
Zhangs-MacBook-Pro:code yue_zhang$ python zero_elif.py
Give me a number: -3
The number is negative
Bye
```

The following program, *scoregrade.py*, further demonstrates a five-way branching example. The program asks the user to enter a final-exam score, and then prints the grade (A/B/C/D) that corresponds to the score. If the score that the user entered is less than 0 or greater than 100, the program prints an error message.

```
score = float(raw_input('Enter a score: '))
if 85 <= score <= 100:
    print 'A'
elif 70 <= score < 85:
    print 'B'
elif 50 <=score < 70:
    print 'C'
elif 0 <= score < 50:
    print 'D'
else:
    print 'Invalid score entered'
```

The control flow diagram for this program is shown in Fig. 4.6, where the five branches from left to right correspond to the score being in the ranges [85, 100], [70, 85), [50, 70), [0, 50) and otherwise, respectively. At one execution of the program, only one branch is executed.

```
Zhangs-MacBook-Pro:code yue_zhang$ python grade.py
Enter a score: 75
B
Zhangs-MacBook-Pro:code yue_zhang$ python grade.py
Enter a score: 100
A
Zhangs-MacBook-Pro:code yue_zhang$ python grade.py
Enter a score: 33
```

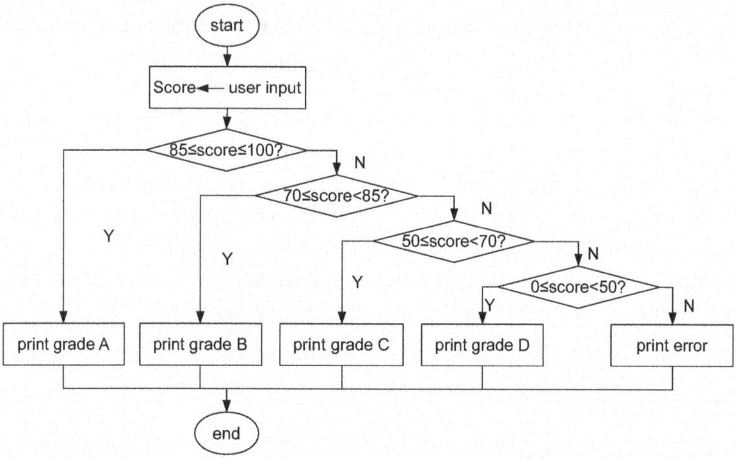

Fig. 4.5 Five-way branching example

```
D
Zhangs-MacBook-Pro:code yue_zhang$ python grade.py
Enter a score: -10
Invalid score entered
```

Note that the program above consists of only two top-level statements: an assignment statement and a compound *if* ... *elif* ... *else* ... statement. The flow chart of a compound *if* ... *elif* ... *else* ... statement is always right-branching—additional *elif* Boolean expressions always split the right branch from its previous branching point.

Finally, the general form of the *if* statement, which consists of one *if* component, zero or many *elif* components, and zero or one *else* component, is shown in Fig. 4.5.

4.2.1 Nested **if** *Statements*

Since a compound *if* statement is also a statement, it can be used in a statement block inside another *if* statement. This form of nested *if* statements is sometimes useful for organizing branches of control flow with complex conditions.

For example, in a FIFA World Cup, four teams participate in the semi-finals by paying two semi-final matches separately. The winners of the two semi-final matches play a final match for the World-Cup championship, while the losers of the two semi-final matches play a match for the third place. In this process, each of the four teams will play two matches, and the results determine the team's final rank in the World-Cup game. The following Python code simulates the game.

Fig. 4.6 General form of the *if* statement

```
[semi_final.py]
first_game = raw_input('Did the team win the first
    game? [Y/N] ')
second_game = raw_input('Did the team win the second
    game? [Y/N] ')
if first_game == 'Y':
    if second_game == 'Y':
        print 'First place'
    else:
        print 'Second place'
else:
    if second_game == 'Y':
        print 'Third place'
    else:
        print 'Fourth place'
```

The program consists of three statements, the first two being assignment statements that inquire the user for the results of the two matches that a team played. The third statement is a compound *if* statement that contains two indented statement blocks, each consisting of another *if* statement. An example of executing *semi_final.py* is shown below.

```
Zhangs-MacBook-Pro:code yue_zhang$ python semi_final.py
Did the team win the first game? [Y/N] Y
Did the team win the second game? [Y/N] N
Second place
Zhangs-MacBook-Pro:code yue_zhang$ python semi_final.py
Did the team win the first game? [Y/N] N
Did the team win the second game? [Y/N] Y
Third place
```

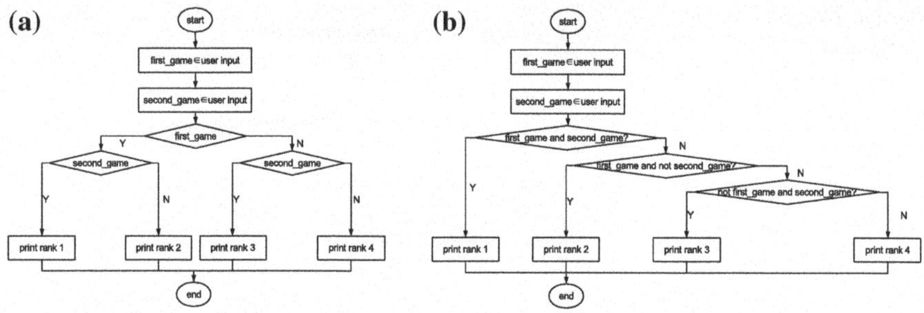

Fig. 4.7 Control flow diagrams. **a** semi_final.py. **b** semi_final_unnested.py

```
Zhangs-MacBook-Pro:code yue_zhang$ python semi_final.py
Did the team win the first game? [Y/N] Y
Did the team win the second game? [Y/N] Y
First place
```

For analyzing the dynamic behavior of nested branching statements, a hierarchical approach can be useful—the branching statement of the outer level can be analyzed first (without considering the behavior of branches in the inner level), before the branching statements in the inner level are analyzed independently. In each level, the *if* statement behaves exactly as introduced earlier depending only on the Boolean expression in the *if* line. Nesting does not change the mechanism of each individual statement. The flow chat of *semi_final.py* is shown in Fig. 4.7a.

An alternative way of writing the same program, without using nested *if* statements, is to use an *if* statement with combined Boolean conditions.

```
[semi_final_unnested.py]
first_game = raw_input('Did the team win the first
   game? [Y/N] ')
second_game = raw_input('Did the team win the second
   game? [Y/N] ')
if first_game and second_game:
   print 'First place'
elif first_game and not second_game:
   print 'Second place'
elif not first_game and second_game:
   print 'Third place'
else:
   print 'Fourth place'
```

An execution of *semi_final_unnested.py* yields the same result as *semi_final.py*. However, the logic behind *semi_final_unnested.py* is different from that behind *semi_final.py*. The control flow diagram of *semi_final_unnested.py* is shown in Fig. 4.7b, where the two matches are combined into one condition, and the four different outcomes are organized in a right-branching hierarchy.

As can be seen from a comparison between Fig. 4.7a, b by using nested *if* statements, multiple Boolean conditions can be organized in an arbitrary hierarchy. In this particular example, nested *if* offers a more balanced and perhaps more intuitive logic hierarchy when compared to the right-branching hierarchy by the compound *if* ... *elif* ... *else* ... statement without using nested *if* statements.

In general, all multi-way branching situations that can be addressed by using *if* ... *elif* ... *else* statements can be addressed by using nested *if* statements also. A choice can be made according to the convenience of expressing a logical hierarchy. Sometimes *if* statements can be nested in more than two levels. When there are many branches and complex hierarchical logic structures, it is possible to make a mistake by writing an incorrect Boolean condition. In such situations, drawing a control flow diagram can be useful for spotting and fixing logical mistakes. In addition, **debugging** techniques, introduced later in this chapter, can be used to verify the behavior of a program dynamically.

4.3 Looping Using the *While* Statement

The looping control flow of Fig. 4.1c can be implemented using **the while statement** in Python. The syntax of the *while* statement is shown in Fig. 4.8. It is identical to the syntax of the *if* statement in Fig. 4.3, except that the keyword is changed from *if* to *while*. For the *while* statement, the indented statement block is also called the **loop body**.

As shown in Fig. 4.1c, the Boolean expression determines the execution of the indented statement block under the *while* line. If it is *True*, the statement block is executed; if it is *False*, the statement block is not executed. This is similar to the *if* statement. However, the *while* statement differs from the *if* statement after the execution of the indented statement block. In particular, after the execution of the indented statement block, the execution of the *if* statement is finished, and any further statements in the program will be executed. In contrast, after the indented statement block is executed, the *while* statement moves back to the Boolean expression in order to evaluate it again. If it is *True* this time, the indented statement block is executed again, before the Boolean expression is evaluated yet again. This process repeats (i.e. *loops*) as long as the value of the Boolean expression is *True*.

For a simple example, consider the following contrast Boolean condition:

```
>>> while False:
...     print 'a'
...
>>>
```

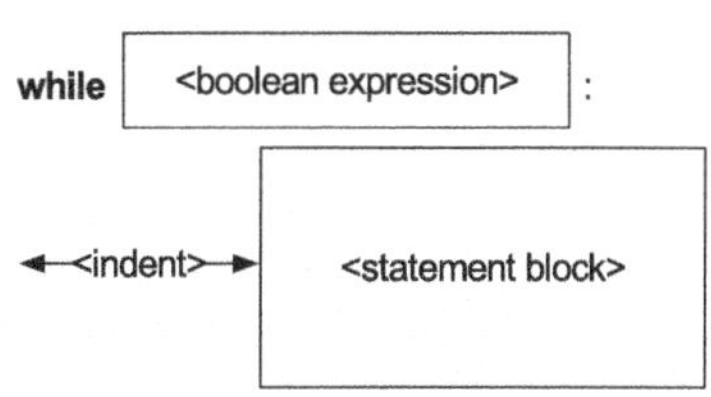

Fig. 4.8 Syntax of the *while* statement

In this example, the *while* statement does not print anything on the console, because the Boolean expression is the constant *False*. This is the same as with the *if* statement. However, when the Boolean expression is turned into *True*, the behavior of the *while* statement is different from the *if* statement:

```
>>> while True:
...     print 'a'
...
a
a
a
a
a
a
a
a
a
a
a
a
a
a
^Ca
a
Traceback (most recent call last):
  File "<stdin>", line 2, in <module>
KeyboardInterrupt
```

In this example, the character 'a' is repeatedly printed. The execution does not stop until an external interrupt happens. In this case, the user stops the execution by using a **keyboard interrupt**, which can be done by pressing and holding the *Ctrl* key and then pressing the *C* key (i.e. pressing *Ctrl-C*). Keyboard interrupts can be used to stop a Python program that is currently being executed.

The infinitely repeating execution of the loop body (i.e. indented statement block in the *while* statement) is called an **infinite loop**, or a **dead loop**. It is one of the most common mistakes that programmers can make by not properly setting the terminating condition of the loop. In order to avoid dead loops, the value of the Boolean expression must be changed to *False* after a finite number of executions of the loop body. An example is shown below.

```
>>> i=0
>>> while i<3:
...     print i
...     i += 1
...
0
1
2
```

The *while* statement above prints the sequence of numbers 0, 1, 2 on the console. The control flow diagram of this statement is shown in Fig. 4.9a. During dynamic execution, each statement in the loop body is executed three times, as illustrated in Fig. 4.9b.

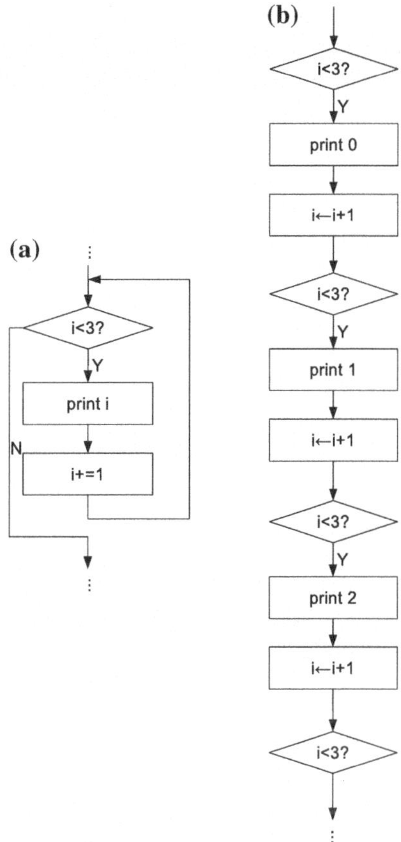

Fig. 4.9 Analysis of the dynamic behavior of a *while* statement. **a** control flow diagram. **b** Dynamic execution sequence

The value of the variable i is set to 0in the beginning of execution. The *while* statement checks the value of i in the beginning of each iteration to decide whether the loop body is executed. Inside the loop body, the value of i is incremented by 1. As a result, each time the Boolean expression $i < 3$ is evaluated, the value of i changes. When the value of i becomes 3 after the third iteration, the Boolean expression $i < 3$ becomes *False* and the loop finishes. It is commen to use the value of a variable (e.g. i in the example above) to control the execution of a loop, and such a variable is also called the **loop variable**.

Another way to control the termination of a loop is via user input. For example, consider a program that generates random letters to a user. The program works by repeatedly generating random letters to the user until she asks it to stop. This can be achieved by using a *while* loop, with the loop body being a random letter generator, and the looping condition being whether the user requested to stop the program.

In order to generate a random letter, one can use a random number generater together with a mapping from numbers to letters. The former is readily available in the *random* module, while the latter can be implemented by using a string that contains all the candidate letters. An integer can be mapped into a letter in the string by serving as the index in a *getitem* operation. Given this random letter generator, the full program can be written as:

```
[lettergen.py]
import random
s='ABCDEFGHIJKLMNOPQRSTUVWXYZ'
command = raw_input('Generate random letter? [Y/N] ')
while command == 'Y':
    i = random.randint(0, len(s)-1)
    print s[i]
    command = raw_input('Generate random letter? [Y/N]
        ')
print 'Bye'
```

The program above consists of five top-level statements. The first is an *import* statement for the *random* module. The second is an assignment statement that defines the string for mapping integers to letters. The string consists of the upper case English alphabet. The third statement is an assignment statement that requests the user to enter a command. The fourth statement is a compound *while* loop, which consists of a statement block that generates a random number and then asks the user for the next command. The fifth statement is a *print* statement that indicates the finishing of execution to the user.

The loop body consists of three statements, the first being an assignment statement that obtains a random integer that ranges over all possible indices on the string, the second being a *print* statement that shows the value of an expression that maps the integer to a letter by using the *getitem* operation, and the third being the assignment statement that asks the user for the next command.

When executed, *lettergen.py* repeatedly asks the user to choose from generating a random letter ('Y') and exiting the program (any input other than 'Y'). Here, the user input is used in the Boolean condition, so that the value of the expression changes at each iteration.

```
Zhangs-MacBook-Pro:code yue_zhang$ python lettergen.py
Generate random letter? [Y/N] Y
K
Generate random letter? [Y/N] Y
J
Generate random letter? [Y/N] Y
Z
Generate random letter? [Y/N] Y
X
Generate random letter? [Y/N] Y
G
Generate random letter? [Y/N] Y
P
Generate random letter? [Y/N] Y
W
Generate random letter? [Y/N] Y
U
Generate random letter? [Y/N] Y
```

```
V
Generate random letter? [Y/N] Y
M
Generate random letter? [Y/N] N
Bye
```

The *while* statement can be extended with an *else* component also, in exactly the same syntax as the *if* statement. The interpretation is the same: when the Boolean expression is evaluated to *False*, the indented statement block under *else* is executed, for once only. For example,

```
>>> while False:
...     print 'a'
... else:
...     print 'b'
...
b
```

The statement above would maintain its behavior if the *while* keyword were changed into *if*. The loop body is not executed at all in this case. In case the loop body is executed a finite number of times, the *else* statement block is executed when the iterations finish (i.e. the first time when the Boolean expression becomes *False*).

```
>>> i=0
>>> while i<3:
...     print i
...     i += 1
... else:
...     print 'Done with i =', i
...
0
1
2
Done with i = 3
```

When the statement above is executed, the dynamic execution sequence starts identical to that of Fig. 4.9. When the value of the loop variable *i* is changed to 3, the Boolean expression is evaluated to *False*. As this happens, the statement block under *else* is executed, with the current value of *i* (i.e. 3) being shown.

In case the loop is infinite, the *else* statement block is never executed.

```
>>> while True:
...     print 'a'
... else:
...     print 'b'
...
a
a
a
a
a
a
a
a
a
```

```
a
a
a
a
a
^Ca
a
Traceback (most recent call last):
  File "<stdin>", line 2, in <module>
KeyboardInterrupt
```

In the example above, the letter ‘b’ is not printed when the execution of the program is interrupted by force, since the Boolean expression is never evaluated to *False*.

4.3.1 Branching Nested in a Loop

As other statements, an *if* statement can be put into the loop body of a *while* statement. For a simple example, suppose that a user needs to print all the integers from 1 to 100 that are not multiples of 3. One solution is to set up a loop variable *i* that counts from 1 too 100, in the same way as the earlier example that counts from 0 to 2, while printing out the value of *i* only when it is not a multiple of 3.

```
[conditional_print.py]
i=1
while i<100:
    if i%3 != 0:
        print i
    i+=1
```

In the example above, the *print* statement is put under the Boolean condition that *i*%3 != 0. Hence in a dynamic execution sequence, it will be executed only when *i* is not a multiple of 3. The analysis of *if* statements nested in *while* loops is similar to the analysis of nested *if* statements—the execution of each statement can be analyzed separately, in an outer-intra hierarchy. Note that the underlying mechanism of each individual *if* and *while* statement, no matter whether nested or not, is the same and relies on the relevant Boolean expression only. Again, a useful technique for analyzing a combination of looping and branching statements is to write a control flow diagram.

For a more complex example, in which the Boolean condition of the loop is affected by the execution of an *if* statement in the loop body, take the following number guessing game. Suppose that the computer generates a random number between 1 and 100, and the user tries to guess it. Every time the user enters a guess, the computer tells whether the guessed number is greater than, equal to, or smaller than the answer. In case the guess is correct, the game ends with the user winning the game. Otherwise the user is allowed to enter another guess. The game repeats until the user finds the answer. An example implementation of the game is as follows:

```
[guess.py]
import random
answer = random.randint(1, 100)
wins = False          # looping condition
while not wins:
    guess = int(raw_input('Your guess is: '))
    if guess == answer:
        print 'You win!'
        wins = True
    elif guess > answer:
        print 'Too large'
    else:
        print 'Too small'
```

guess.py uses the Boolean variable *wins* to record whether the user has won the game. The variable is initialized to *False*. Until *wins* is changed to *True*, the program repeats asking the user for a guess, and giving relevant feedback to the user. *wins* is set to *True* if the answer at the current iteration is correct. An example execution of *guess.py* is shown below.

```
Zhangs-MacBook-Pro:code yue_zhang$ python guess.py
Your guess is: 50
Too large
Your guess is: 25
Too small
Your guess is: 37
Too small
Your guess is: 43
Too large
Your guess is: 40
Too small
Your guess is: 42
Too large
Your guess is: 41
You win!
```

The program above can be made more challenging by setting a limit on the number of guesses allowed.

```
[guess_limit.py]
import random
limit=7
answer = random.randint(1, 100)
guesses=0
wins = False
while guesses < limit and not wins:
    guess = int(raw_input('Your guess is: '))
    guesses += 1
    if guess == answer:
        print 'You win!'
        wins = True
    else:
        if guess > answer:
            print 'Too large'
        else:
            print 'Too small'
```

```
else:
    if not wins:
        print 'You lose.'
```

guess_limit.py uses an additional variable, *guesses*, to count the number of guesses that the user has already made. An extra condition that *guesses* is smaller than *limit* is added to the Boolean expression for *while*, which decides whether to allow the user to guess again. In addition, when the loop finishes, either *wins* is *True* or *guesses* is equal to *limit*. In the latter case, the user has lost the game. A corresponding *else* statement block is added so that a message is shown to the user the the game finishes without being won.

Loops can also be put in other loops, resulting a **nested loops**, more discussions of which is given in Chap. 9.

4.3.2 Break and Continue

break and **continue** are two statements that are associated with loops; they must be put in a loop body. When *break* is executed, Python finishes the loop by directly jumping out of the looping statement (e.g. *while*), executing any successing statements. When *continue* is executed, Python finishes the current loop iteration without executing the rest of the loop body, jumping back to the Boolean condition again. For example,

```
>>> i=1
>>> while i<3:
...     print i
...     break       # jump out directly
...     i += 1
... else:
...     print 'done'
...
1
>>> print i
1
```

In the example above, the statement *print i* in the *while* loop is executed only once, because the first time the *break* statement is executed, the loop finishes. The final value of *i* is 1, since $i+=1$ is not executed. Because the Boolean expression $i < 3$ is not evaluated a second time, it does not have a chance to become *False*, and therefore the *else* statement block is not executed.

An example for *continue* is shown below.

```
>>> i=1
>>> while i<3:
...     i += 1
...     continue    # jump to while
...     print i
... else:
...     print 'done'
```

```
...
done
>>> print i
3
```

In the example above, the statement *print i* in the *while* loop is never executed. This is because every time *continue* is executed, the rest of the loop body is skipped. However, the loop continues until the Boolean expression $i < 3$ is changed to *False*, as demonstrated by the 'done' message and the final value of i.

The use of *break* and *continue* can sometimes make a program more intuitive to understand. A version of *guess_break.py* that uses *break* instead of a complex Boolean condition is:

```
[guess_break.py]
import random
answer = random.randint(1, 100)
limit = 7
guesses = 0
while guesses < limit:
    guess = int(raw_input('Your guess is: '))
    guesses += 1
    if guess == answer:
        print 'You win!'
        break
    else:
        if guess > answer:
            print 'Too large'
        else:
            print 'Too small'
else:
    print 'You lose.'
```

The main structure of *guess_break.py* is a loop that repeats for *limit* iterations, asking the user for a guess at each iteration. If the guess is correct, it prints the winning message and terminates the loop. Otherwise it prints out the corresponding hint message and the loop continues. If the loop finishes when the Boolean expression *guesses < limit* becomes *False*, the *else* statement block is executed, printing out the losing message.

4.4 Debugging

With branching and looping statement, the dynamic execution sequence of a program can be very different from the static program itself. Programming mistakes can happen when the Boolean conditions for loops and branches are incorrect, or certain factors in the dynamic control flow are overloaded. In order to find potential errors in a program, it is useful to trace the dynamic execution of the program, monitoring the values of variables in the process, and this task is also **debugging**. There are two common debugging techniques: tracing and asserting.

Tracing refers to the monitoring of variable values in a dynamic execution process. It helps to understand the dynamic execution sequence of a piece of code. For example, suppose that *guess_break.py* intended to allow the user to enter up to 8 guesses. However, execution of the program only allows 7 guesses. In order to figure out the dynamic execution of the code, on can trace the values of *guesses* and *limit* in each loop iteration by adding a *print* statement.

```
[guess_break.py]
import random
answer = random.randint(1, 100)
limit = 7
guesses = 0
while guesses < limit:
    print 'TRACE:', guesses, limit
    guess = int(raw_input('Your guess is: '))
    guesses += 1
    if guess == answer:
        print 'You win!'
        break
    else:
        if guess > answer:
            print 'Too large'
        else:
            print 'Too small'
else:
    print 'You lose.'
```

An execution of the program can lead to the following output:

```
TRACE: 0 7
Your guess is: 50
Too large
TRACE: 1 7
Your guess is: 25
Too large
TRACE: 2 7
Your guess is: 13
Too small
TRACE: 3 7
Your guess is: 17
Too small
TRACE: 4 7
Your guess is: 20
Too small
TRACE: 5 7
Your guess is: 23
Too large
TRACE: 6 7
Your guess is: 21
You lose!
```

The trace information shows that the value of *guesses* increases from 0 to 6 in 7 dynamic interactions. A further check of the looping condition reveals that the problem is that the value becomes 7 after the last iteration, dissatisfying $guesses < limit$. This can be fixed by setting the Boolean condition to $guesses \leq limit$.

As another example of tracing, *conditional_print.py*, introduced earlier in this chapter, can be debugged by printing out the values of *i* at each iteration, ensuring the correct range of the loop variable.

```
[conditional_print.py]
i=1
while i<100:
    print 'TRACE:i=%d'%i
    if i%3==0:
        print i
    i+=1
```

After debugging, the trace print can be deleted or commented out.

Asserting is a special command to ensure that certain condition is true at a special point of execution. In Python, asserting is achieve by the **assert statement**, the syntax of which is

```
    assert <Boolean expression>
```

When an *assert* statement is executed, if the Boolean expression is evaluated to *True*, no operations are taken. If the Boolean expression is evaluated to *False*, an error will be raised.

```
>>> assert True
>>> assert False
Traceback (most recent call last):
  File "<stdin>", line 1, in <module>
AssertionError
```

The *assert* statement can be used to check for design assumptions during debugging. If some condition is assumed to be true during dynamic execution, it can be put into an assert statement. If it is *False* in any execution of the *assert* statement, the execution halts with an *assertion error*, through which the programmer cab obtain certain clue about coding errors. The *assert* statement can be used in *guess_break.py* to ensure that the last *else* statement block is executed only when *guess* < *answer*, a condition that the program design implies, but is not explicitly written in the *else* statement. Asserting this condition can avoid unexpected errors during debugging executions.

```
[guess_break.py]
import random
answer = random.randint(1, 100)
limit = 7
guesses = 0
while guesses < limit:
    guess = int(raw_input('Your guess is: '))
    guesses += 1
    if guess == answer:
        print 'You win!'
        break
    else:
        if guess > answer:
            print 'Too large'
        else:
```

```
        assert guess < answer
        print 'Too small'
else:
    print 'You lose.'
```

After debugging, the *assert* statement can be deleted or commented out from a program, because it is not a part of the main functionality.

Using IDLE for debugging. For many programming languages, special software applications are available for debugging, and such applications are called **debugging tools**. The basic idea of a debugging tool is to control the runtime, so that it is possible to freeze the dynamic execution of a program at any point, and inspect the values of variables at that point. In other words, debugging tools offer convenience for tracing. In order to achieve this, the program to debug must be executed from the debugging tool.

Two typical functionalities of a debugging tool are **break points** and **stepping**. *Break points* can be placed by a debugging tool in a statement in a program. During an execution of the program from the debugging tool, whenever a break point is hit, the execution pauses, and the debugging tool shows the current values of all variables. *Stepping* allows the user to execute only the next statement in a program, before the program is frozen again by the debugging tool.

Figure 4.10 shows the IDLE debugging tool for Python, which displays the program to be debugged. The debugging tool can be loaded from IDLE by using Run→Python Shell→Debug→Debugger→Run Module. To set a break point before a statement, first select the statement, second click the right button, and then choose 'Set Breakpoint', as shown Fig. 4.11. As an example, a breakpoint can be set before line 6 of *guess_break.py*, so that the execution is frozen every time the loop body is executed. When the program is frozen, the values of *guesses* and *limit* can be inspected in Fig. 4.10, in order to ensure that the loop terminates after the given limit. Stepping can be performed by clicking the *step* button on the top of Fig. 4.10.

As mentioned earlier in this chapter, the sequential, branching and looping structures in Fig. 4.1 are the three basic types of control flow. All problems that are solvable by a computer can be addressed by using a combination of the three basic types of dynamic structures. Understanding the dynamic behavior is the beginning of becoming a programmer. However, given a problem, how to design a program that gives the most effective solution is another important question. The next chapter discusses two typical problems and their basic solutions.

Exercises

1. Write a program that

    (a) asks the user for two integers a and b, and prints 'a' if $a > b$, 'ab' if $a = b$, and 'b' if $a < b$.

    (b) prints the string 'Hello' 123 times, each in a line.

    (c) asks the user for three integers a, b and c, and prints out the largest among them. If there are more than one largest value (e.g. a=b and a>c), print 'Tie'.

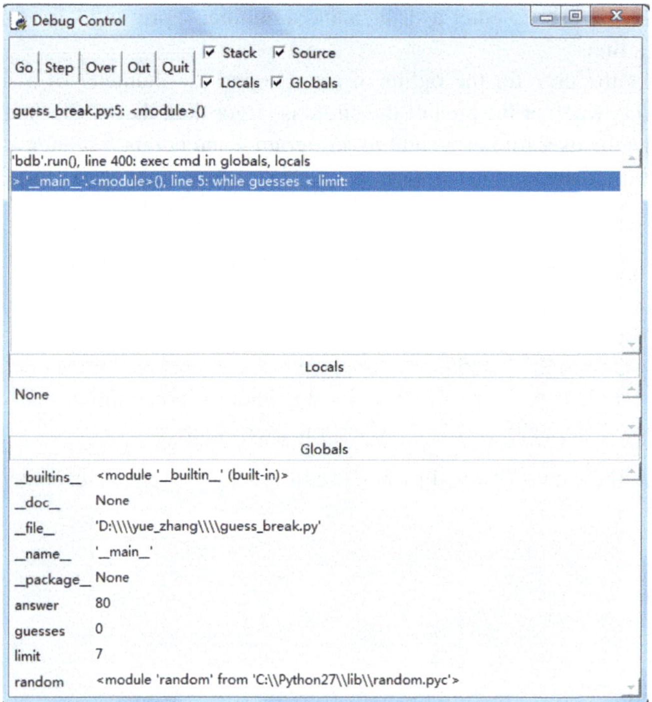

Fig. 4.10 Debugging tool for Python

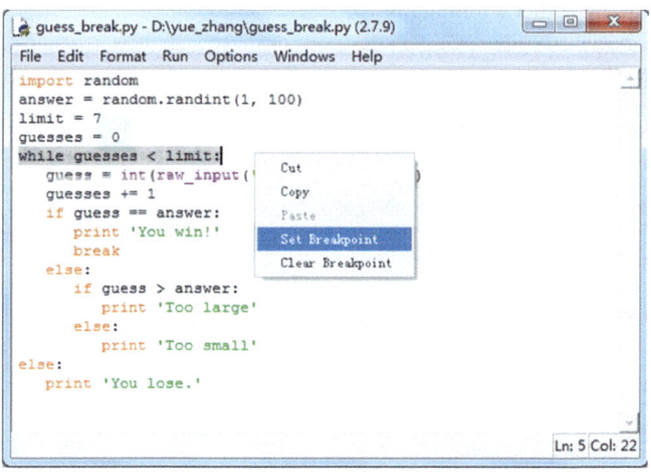

Fig. 4.11 Setting breakpoint

(d) asks the user to enter a string s and a number i, printing s for i times, each in a line.
(e) asks the user for the radius of a circle and the diameter of a square, and prints whether the area of the circle is larger than that of the square.
(f) asks the user for her weight w (kilograms) and height h (meters), and prints whether she is fit ($18.5 \leq w/h^2 \leq 25$), underweight or overweight.

2. Write a program that prints the following sequence of numbers:

   (a) 3, 6, 9, 12, ..., 150.
   (b) 1.0, 0.7, 0.4, 0.1, −0.2, ..., −5.0.
   (c) 3, 9, 27, ..., 3^{10}.
   (d) 1, −2, 4, −8, ..., 256.
   (e) 1, −1/3, 1/5, ..., $(-1)^{i+1}/(2i-1)$, where i is user input.
   (f) 1/0!, 1/1!, 1/2!, ..., $1/i!$, where i is user input.

3. What are the values of s and i when the following program is executed?

```
s=0
i=0
n=10
while i<n:
    if i>n/3:
        s+=i
    i+=1
#print i,s
```

4. Which of the following programs are incorrect? Why?

   (a) decide whether a number is 10.

```
i = int(raw_input("i="))
if i = 10:
    print 'yes'
else:
    print 'no'
```

   (b) decide whether a number is negative.

```
i = int(raw_input("i="))
if i > 0:
    print 'positive'
else:
    print 'negative'
```

   (c) print from 1 to 10.

```
i=1
while i != 10:
    print i
```

   (d) print 1, 3, 5, 7, 9

```
i=1
while i != 10:
    print i
    i += 2
```

(e) print 1, 3, 5, 7, 9

```
i=1
while i<10:
    print i
    if i % 2 == 0:
        continue
    i += 2
```

(f) print 1, 2, 3

```
i=1
while i<10:
    print i
    if i==4:
        continue
    i += 1
```

5. Write a program that translates a data written in the DD-MM-YY format into the Month Day, Year format. For example, an execution of the program could be:

```
Specify a date: 050980
The date is: September 5th, 1980
```

In the example above, the program prompts the message "Specify a date: " and asks the user to input a date. The user enters "050980", and the program output "September 5th, 1980". Hint: consider how two slice a split to get MM, DD and YY, and how to map a number to a month name (January, . . ., December). Use *if* to decide whether "st', "'nd", "rd" or "th" should be added to the back of a date (e.g. 21st, 22nd, 23rd), and whether "19" and "20" should be added to the front of a year (e.g. 1980, 2010).
6. What are two common techniques in debugging? How can they be applied to Python code? Debug the programs in questions 3 and 4 with and without using the IDLE debugger.

Chapter 5
Problem Solving Using Branches and Loops

This chapter discusses the solutions to several common types of problems, showing how typical methods can be developed using branches and loops. It starts with a set of basic and abstract problems, showing how they are correlated to the basic problem of summation. Then it discusses a set of numerical analysis problems, which study approximate answers to mathematical problems that do not have an exact solution. The basic problem solving techniques are applied to numerical integration, root finding and differential equations. Finally, this chapter introduces the tuple type in Python, a container type that represents a sequence of objects, showing how the basic problems such as summation can be performed on tuples by traversal.

5.1 Basic Problems

As mentioned in the previous chapter, all solvable problems can be approached by using a combination of branches and loops. Understanding the dynamic control flow is the first step towards problem solving using programming, but the ultimate purpose is to combine and use basic control flows for solving problems. This chapter introduces several typical problems and discusses their solutions.

5.1.1 Summation

One of the most basic problems that can be solved by using loops is the problem of **summation**, which can be formulated as $s = \sum_{i=m}^{n} f(i)$, where m and n are two integers, and $m \leq n$. The simplest form of summation is $s = \sum_{i=m}^{n} i$, which can be solved by the following program.

```
[sum.py]
m=0         # the start
n=100       # the end
s=0         # the sum
```


Y. Zhang, *An Introduction to Python and Computer Programming*,
Lecture Notes in Electrical Engineering 353, DOI 10.1007/978-981-287-609-6_5

```
i=m        # the iterator
while i<=n:
    s += i
    i += 1
print 'The sum of integers from %d to %d is %d' % (m, n, s)
```

sum.py calculates the sum by accumulating its value incrementally. It uses two variables, i and s, to represent the current value to be added to the sum and the sum up to the current iteration, respectively. While s is initialized to 0, i is initialized to the beginning value m. The *while* statement uses i as the loop variable, which increases from m to n by a step size of 1. At each iteration, the value of i is added to the sum s, and then increased by 1. The key idea behind this program is the use of a sum value, which is incrementally updated in an iterative process. This basic way of thinking applies to many other problems, as will be shown throughout this book. Executing *sum.py* yields the following output.

```
Zhangs-MacBook-Pro:code yue_zhang$ python sum.py
The sum of integers from 0 to 100 is 5050
```

For a more complex example, consider the summation of all multiples of 3 from 1 to 100. There are many different ways to solve this problem. One solution is to iterate overall all multiples of 3, adding them to a sum value.

```
[sum3.py]
m=0       # the start
n=100     # the end
s=0       # the sum
i=0       # the iterator
while i<=100:
    s += i
    i += 3
print 'The sum of all multiples of 3 from 0 to 100 is', s
```

In contrast to *sum.py*, *sum3.py* takes a different strategy to perform the iteration: the initial value of i is set to 3 rather than 1, and the step size is also set to 3. In this way the values of i cover all multiples of 3 from 1 to 100. The other aspects of *sum3.py* are the same as *sum.py*. An execution of *sum3.py* yields the following output.

```
Zhangs-MacBook-Pro:code yue_zhang$ python sum3.py
The sum of all multiples of 3 from 1 to 100 is 1683
```

An alternative solution to the problem above is to compute $\sum_{i=1}^{100/3} 3i$.

```
[sum3_f.py]
s=0
i=1
while i<=100/3:
    s+=3*i
    i+=1
print 'The sum of all multiples of 3 from 1 to 100 is', s
```

sum3_f.py iterates over the integers from 1 to 100/3, which fall between 1 and 100 when multiplied by 3. At each iteration, it accumulates the value of $f(i) = 3i$

into the sum. The execution of *sum3_f.py* gives the same outputs as the execution of *sum3.py*

Production. The method of summation can be generalized to solve the problem of **production** by changing the addition operation to the multiplication operation. For example, the factorial function $n!$ can be calculated by the following program.

```
[factorial.py]
n=10
s=1
i=1
while i<=n:
    s *= i
    i += 1
print '%d! = %d' % (n, s)
```

factorial.py follows the same strategy as *sum.py*, with the product s being initialized to 1 instead of 0, and the incremental operator being changed from += to *=. Execution of the program yields the following result.

```
Zhangs-MacBook-Pro:code yue_zhang$ python factorial.py
10! = 3628800
```

Maximum. The problem of summation can also be generalized into the tasks of finding the **maximum** and **minimum** values of a function $f(x)$, which can be formulated as $\mathbf{max}_{i=m}^{n} f(x)$ and $\mathbf{min}_{i=m}^{n} f(x)$, respectively. Take finding the maximum value for example. Again one solution can be based on *sum.py*, with the initial value of s being set to the negative infinity instead of 1, and the incremental operator being *max* rather than +=.

```
[max.py]
import math
m=0
n=100
s=float('-inf')
i=m
while i<=n:
    f = math.sin(i*i)
    s = max(f, s)
    i += 1
print 'The maximum value of the function is', s
```

max.py finds the maximum value of $sin(x^2)$, where x is an integer that ranges from 0 to 100. It has a similar structure as *sum.py* except for the aforementioned changes to the initial value of s and the incremental operator. Here s is used to represent the current maximum value of f at each step. When the iterations finish, s will be the global maximum. The *math* module is used to calculate the value of $sin(x^2)$, and the expression *float('-inf')* is used to obtain the **negative infinity**. The *max* function is a built-in function of Python; it can take two or more numerical arguments, and gives the maximum of its input arguments as the return value. An execution of *max.py* yields the following result.

```
Zhangs-MacBook-Pro:code yue_zhang$ python max.py
The maximum value of the function is 0.999765729024
```

A common alternative implementation of *max.py* uses an *if* condition instead of the *max* function in the incremental step.

```
[max_if.py]
import math
m=0
n=100
s=float('-inf')
i=m
while i<=n:
    f = math.sin(i*i)
    if f>s:
        s=f
    i += 1
print 'The maximum value of the function is', s
```

The *if* statement achieves the same purpose as the *max* function: it updates the current maximum value if the current value of f is higher than the maximum value in the previous iterations. *max_if.py* yields the same result as *max.py*. One advantage of this program is that it is easy to obtain the value of i that gives the maximum value of f, in addition to the maximum value of f itself.

```
[argmax_if.py]
import math
m=0
n=100
s=float('-inf')
i=m
argmax=None
while i<=n:
    f = math.sin(i*i)
    if f>s:
        s=f
        argmax=i
    i += 1
print 'The maximum value of the function is', s, 'achieved
    when the input is', argmax
```

argmax_if.py uses an additional variable, *argmax*, to store the value of i that corresponds to the maximum value of f. The initial value of *argmax* is not important. Whenever the value of s is updated, the value of *argmax* is also updated. An execution of the program gives the following output.

```
Zhangs-MacBook-Pro:code yue_zhang$ python argmax_if.py
The maximum value of the function is 0.999765729024
   achieved when the input is 75
```

Decision. Interestingly, the method of summation can also be generalized to solve **the decision problem** over a sequence of numbers. One example of such decision problems is to decide whether there is an integer root of $x^4 - 93x - 19620 = 0$ between 1 and 20. The problem can be formulated as $\mathbf{or}_{i=m}^{n}(f(i) == 0)$. A solution can be based on *sum.py*, with the initial value of s being set to *False* instead of 1, and the incremental operation being set to *or* instead of +=.

```
[root.py]
m=1
n=20
s=False
i=m
while i<=n:
    s = s or (i*i*i*i-93*i-19620==0)
    i += 1
print 'The fact that there is a root is', s
```

An execution of the program above yields the following output.

```
Zhangs-MacBook-Pro:code yue_zhang$ python root.py
The fact that there is a root is True
```

Useful in some situations, the idea behind *root.py*, which incrementally updates a Boolean answer, might not be the most intuitive solution to the decision problem. To change the perspective, deciding whether there is a root of $f(x)$ between m and n is essentially the same as deciding whether the user finds the correct guess within l iterations in the number guessing game in the previous chapter. A solution can be derived from *guess_break.py*.

```
[root_break.py]
m=1
n=20
i=m
while i<=n:
    if i*i*i*i-93*i-19620==0:
        print 'There is a root of the function'
        break
    i += 1
else:
    print 'There is not a root of the function'
```

root_break.py iterates from m to n, checking whether $i^4 - 93i - 19620 = 0$ at each iteration. If the condition is met at any iteration, the loop stops with the positive answer being displayed. If the loop finishes without the condition being satisfied, the negative answer is displayed. The solution above is essentially to iterate through a set of candidates, searching for a candidate that satisfies a condition. This is a basic example of **the search problem**.

Counting. Another problem that is related to the problem of summation is **counting**. The simplest form of counting can be regarded as $\sum_{i=m}^{n} 1$, a special case of summation: accumulating 1 on every iteration. In many situations, counting is conditional, and can be formulated as $\sum_{i=m}^{n} count(i)$, where *count*(i) can be 0 or 1, based on a specific condition. This form of the *count* function can be implemented using an *if* statement. The following program counts the number of integers x between 1 and 100 for which the function $sin(x^3)$ is positive.

```
[count_fun.py]
import math
m=1
n=100
```

```
s=0
i=m
while i<=n:
    if math.sin(i*i*i)>0:
        s += 1
    i += 1
print 'The number of integers between %d and %d for
    which the function is positive is %d' % (m,n,s)
```

An execution of the program above yields the following output.

```
The number of integers between 1 and 100 for which the
    function is positive is 60
```

5.1.2 Iteratively Calculating Number Sequences

Up to this point in the chapter, a set of example problems on summation and its generalizations has been discussed. Another typical problem that can be solved by using loops is the problem of **calculating number sequences**, which is mathematically different from the summation problem, but can also be solved using an iterative approach. To some readers, it might be more intuitive to calculate number sequences by iteration than to perform summation by iteration.

The typical problem is to calculate the value of a number is a sequence, in which each element depends on a finite number of its predecessors. A simple example is as follows.

$$x_0 = x$$
$$x_{i+1} = f(x_i)$$

In the equations above, x is an **initial value**, and $x_{i+1} = f(x_i)$ is a **difference equation** that represents the relation between two consecutive elements in the sequence. The different equation is simple because x_{i+1} is dependent only on x_i, but not x_{i-1} or other values. For a practical case, consider the problem of bank accounts again. Suppose that an initial sum of money $x_0 = x$ is deposited into a savings account, and no further transactions are made to that account. If the annual interest rate is α, the sequence of numbers that represent the balances after n consecutive years can be written as:

$$x_{i+1} = x_i \cdot (1 + \alpha)\ (1 \leq i \leq n)$$

As shown in Chap. 3, for the savings problem, each element in the sequence above has a closed-form representation, $x_n = x_0 \cdot (1 + \alpha)^n$, and therefore the problem of finding x_n given x_0 can be solved by a simple Python expression (c.f. *savings.py* in Chap. 3). However, for the convenience of illustration, the following program shows how x_n can be calculated using an iterative approach instead.

```
[savings_iter.py]
print '[Iterative Savings Calculator]'
rate = float(raw_input("Please enter the interest rate:"))
years = int(raw_input("Please enter the number of years:"))
init = float(raw_input("Please enter the inital sum of
    money: $"))
i=0
xi=init
while i<years:
   xi = xi * (1+rate)
   i += 1
final = xi
print "After %d years, the savings will be $%.2f" % (years,
     final)
```

An execution of *savings_iter.py* yields the same result as *savings.py* in Chap. 3.

```
[Iterative savings Calculator]
Please enter the interest rate: 0.036
Please enter the number of years: 5
Please enter the inital sum of money: $10000
After 5 years, the savings will be $11934.35
```

savings_iter.py uses the variable i as the loop variable, which represents the current iteration, and the variable xi to represent the value of x_i in the sequence. xi is initialized to the first element x_0, and updated iteratively in each iteration according to the difference equation. In contrast to *savings.py*, which calculates a single value x_n based on the initial value x_0, *savings_iter.py* calculates the whole sequence of numbers $x_0, x_1, \ldots, x_n$ by taking n iterations. Despite of this, only one variable xi is used in the whole process, dynamically representing the ith element x_i at each iteration.

The same technique can be used for induction on more complex series of numbers. For example, a famous number sequence, the Fibonacci numbers, can be formulated as follows:

$$
\begin{aligned}
f_0 &= 1 \\
f_1 &= 1 \\
f_i &= f_{i-1} + f_{i-2}, i > 1
\end{aligned}
$$

The Fibonacci numbers are defined over two initial values f_0 and f_1, and a difference equation. The iterative approach can be used to induce the value of f_n from the values of f_0 and f_1. This time, however, one variable xi is insufficient for representing the whole sequence of numbers. This is because the different equation calculates the value of f_i based on its two predecessors f_{i-1} and f_{i-2}, rather than its immediate predecessor f_{i-1} alone. As a result, at least two variables are needed to record the history at any iteration.

It turns out to be sufficient to use two variables, xi and xii, to dynamically represent the pair of numbers f_i and f_{i-1} at each iteration, respectively. At each step, the new value of xi is the sum of the old values of xi and xii, while the new value of xii is the old value of xi. Formally,

$$f_i^{(new)} \leftarrow f_i^{(old)} + f_{i-1}^{(old)}$$

$$f_{i-1}^{(new)} \leftarrow f_i^{(old)}$$

Note, however, that the two changes must be applied simultaneously, because if the value of xi is updated first, then the old value $f_i^{(old)}$ is no longer available for the updating of xii. Vice versa, if the value of xii is updated first, then its old value is no longer available for the updating of xi.

One typical solution to the problem above is to use a temporary variable, t, to hold a copy of the old value of xi, so that xi can be updated first, before xii is updated using the value stored in t:

$$\begin{aligned} &t \leftarrow f_i^{(old)} \\ &f_i^{(new)} \leftarrow f_i^{(old)} + f_{i-1}^{(old)} \\ &f_{i-1}^{(new)} \leftarrow t \end{aligned}$$

The three changes can now be made sequential, and translated into three Python statements.

```
t=xi
xi=xi+xii
xii=t
```

Given the observation above, the Fibonacci numbers can be calculated iteratively using the following program.

```
[fib_iter.py]
n=10
i=1
xi=1  # f1
xii=1 # f0
while i<n:
   t=xi
   xi=xi+xii
   xii=t
   i += 1
print "The %dth Fibonacci number is %d" % (n, xi)
```

fib_iter.py uses the variable i as the loop variable, and iteratively updates the values of xi and xii. The variables xi and xii are initialized to f_1 and f_0, respectively, and updated according to the difference equation. After $n-1$ iterations, the value of xi represents the value of f_n. An execution of *fib_iter.py* yields the following output.

```
Zhangs-MacBook-Pro:code yue_zhang$ python fib_iter.py
The 10th Fibonacci number is 89
```

5.2 Numerical Analysis Problems

Numerical analysis is an important task that can be solved by computer programs. Although many mathematical problems can be solved by using Python operators and functions, there is a wide range of arithmetic problems for which there is no closed-form solution. No direct operators or library function calls are available for such problems. Examples of this type of problems include root finding, differentiation, integration and differential equations, for which the solution may not correspond to any pre-defined mathematical function. There are also problems for which the exact answer is unknown, but can only be estimated to a certain precision. The value of π is such an example.

The problems are typically solved by computers using numerical methods, which can approximate the real answer to a desired degree of precision. In fact, many functions in the *math* module, including *sin*, *cos* and *log*, are implemented using numerical methods. This section discusses several typical numerical problems, building solutions based on the approaches discussed in the previous section.

5.2.1 Numerical Differentiation

To begin with, consider the task of **differentiation**, which is to find the rate of change of a function at a particular point. The problem is illustrated in Fig. 5.1. The definition of $f'(x_0)$ is the value of $(f(x_0+d)-f(x_0))/d$ when d tends to 0. For some functions, such as $f(x) = x^2$, there is a closed-form derivative (i.e. $f'(x) = 2x$), and hence a direct answer to the differentiation problem can be obtained by using simple operators or pre-defined functions.

For a function $f(x)$ that does not have a known derivative function $f'(x)$, $f'(x_0)$ can be approximated to a certain degree of precision by using $(f(x_0+d)-f(x_0))/d$,

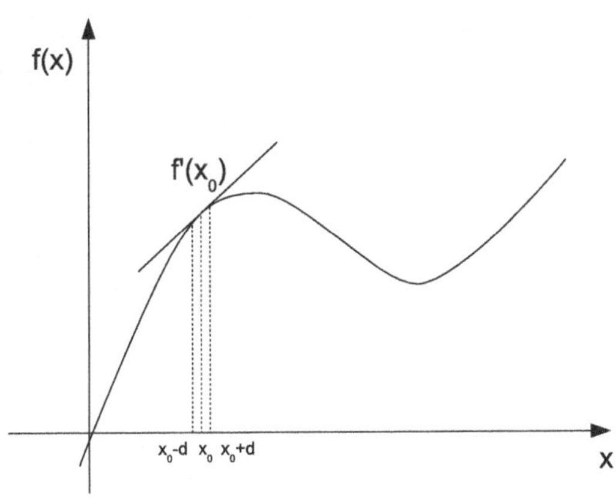

Fig. 5.1 Numerical differentiation

with a value d that is sufficiently small. For example, to calculate the derivative of x^2 at 3.5, the following code can be used.

```
>>> d0=0.001
>>> x0=3.5
>>> ((x0+d0)*(x0+d0)-x0*x0)/d0
7.000999999998925
```

A smaller value of d can be used to solve the same problem.

```
>>> d1=0.000001
>>> x0=3.5
>>> ((x0+d1)*(x0+d1)-x0*x0)/d1
7.000001000179168
```

In the example above, two values of d are used, including 10^{-3} and 10^{-6}, which give different approximations to the true derivative. Since the function $f(x) = x^2$ has a closed-form derivative $f'(x) = 2x$, the true derivative can be calculated as $2x_0 = 7.0$. The numerical answers using $d = 10^{-3}$ and 10^{-6} approximates this true answer by reasonably small errors, while a smaller d gives a result that is closer to the real answer.

5.2.2 Numerical Integration

As illustrated in Fig. 5.2a, the integral $\int_a^b f(x)dx$ can be calculated numerically by dividing the span between a and b into n equal intervals, and summing up the areas under $f(x)$ for each interval. Denoting the width of each interval with d, it holds that $d = \frac{b-a}{n}$. The integral can be calculated as:

$$\int_a^b f(x)dx \approx \sum_{i=0}^{n-1} area_i,$$

where $area(i)$ represents the area of the ith interval, $0 \leq i < n$. As indicated by the equation, the problem of **integration** can be solved numerically using summation.

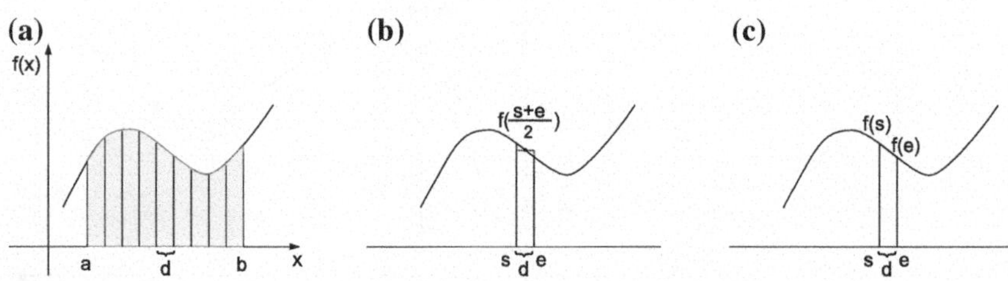

Fig. 5.2 Numerical integration. **a** Numerical integration. **b** Mid-point rule. **c** Trapezoidal rule

There are several ways to calculate the size of each area. One way, as illustrated by Fig. 5.2b, is to treat the area as a rectangle, and use the value of the function at the mid-point (i.e. $(s+e)/2$) as the height. The size of the area under this assumption is $M = d \cdot f((s+e)/2)$. Because the area is estimated by using the value of the function at the mid-point of the interval, the method is called the *mid-point rule*.

Denoting the beginning of the ith interval between a and b as $a_i = a + id$, and the end of the ith interval between a and b as $b_i = a + (i+1)d$, the mid-point of the ith interval between a and b can be written as $(a_i + b_i)/2 = a + (i + 0.5)d$. Therefore, the ith area $area_i$ can be calculated by

$$area_i = d \cdot f(\frac{a_i + b_i}{2}) = d \cdot f(a + (i + \frac{1}{2})d)$$

Therefore, the integral $\int_a^b f(x)dx$ can be calculated by:

$$\int_a^b f(x)dx \approx \sum_{i=0}^{n-1} d \cdot f(a + (i + \frac{1}{2})d) = d \cdot \sum_{i=0}^{n-1} f(a + (i + \frac{1}{2})d)$$

An alternative way of calculating the size of each sub area, as illustrated in Fig. 5.2c, is to treat the area as a trapezoid, with the height on the left being $f(s)$ and the height on the right being $f(e)$. The size of the trapezoid is $T = \frac{(e-s)(f(s)+f(e))}{2}$. Because the area is calculated as a trapezoid, the method is called the *trapezoidal rule*.

Again, denoting the beginning of the ith interval between a and b as $a_i = a + id$, and the end of the ith interval between a and b as $b_i = a + (i+1)d$, the area that correspond to the ith interval can be calculated as:

$$area_i = \frac{d \cdot (f(a+id) + f(a+(i+1)d))}{2} = d \cdot \frac{f(a+id) + f(a+(i+1)d)}{2}$$

And the integral $\int_a^b f(x)dx$ can be calculated as:

$$\begin{aligned}\int_a^b &\approx \textstyle\sum_{i=0}^{n-1} d \cdot \frac{f(a+id)+f(a+(i+1)d)}{2} \\ &= d \cdot \textstyle\sum_{i=0}^{n-1} \left(\frac{f(a+id)}{2} + \frac{f(a+(i+1)d)}{2}\right) \\ &= d \cdot \left(\frac{f(a)}{2} + \textstyle\sum_{i=1}^{n-1} f(a+id) + \frac{f(b)}{2}\right) \\ &= d \cdot \left(\frac{f(a)+f(b)}{2} + \textstyle\sum_{i=1}^{n-1} f(a+id)\right)\end{aligned}$$

It can be shown that as the value of d decreases, the errors of the numerical approximations using both the mid-point rule and the trapezoidal rule decrease. On the other hand, their errors complements each other. To take advantage of this fact, if $\frac{2M+T}{3}$ is used to calculate the size of each subarea instead of M or T, the error can be significantly smaller. This method is called the *Simpson's rule*. Using this rule, the integral $\int_a^b f(x)dx$ can be calculated as:

$$\int\limits_a^b f(x)dx \approx \sum_{i=0}^{n-1} \frac{2df((a+(i+\frac{1}{2})d))+d\frac{f(a+id)+f(a+(i+1)d)}{2}}{3}$$

$$= \frac{d}{6}\big(f(a) + 2\sum_{i=1}^{n-1} f(a+id) + 4\sum_{i=0}^{n-1} f(a+(i+\tfrac{1}{2})d) + f(b)\big)$$

A program that calculates the approximation of $\int_a^b f(x)dx$ using Simpson's rule can be based on the technique of summation, calculating the two terms $2\sum_{i=1}^{n-1} f(a+id)$ and $4\sum_{i=0}^{n-1} f(a+(i+\frac{1}{2}d)$ separately.

```
[simpson.py]
a=0
b=10
n=10000
d=float(b-a)/n
#f(a)
fa=a*a
#f(b)
fb=b*b
# sum f(a+id)
s1=0
i=1
while i<=n-1:
    s1 += (a+i*d) * (a+i*d)
    i += 1
# sum f(a+(i+0.5)d)
s2=0
i=0
while i<=n-1:
    s2 += (a+(i+0.5)*d) * (a+(i+0.5)*d)
    i += 1
# area
s = (d/6) * (fa+2*s1+4*s2+fb)
print 'The integral is equal to', s
```

simpson.py calculates the integral $\int_0^{10} f(x)dx$ for $f(x) = x^2$ according to Simpson's rule. The number of sub areas is set to 10^4. The program breaks the calculation of the integral into four terms, $f(a)$, $2\sum_{i=1}^{n-1} f(a+id)$, $4\sum_{i=0}^{n-1} f(a+(i+\frac{1}{2})d)$ and $f(b)$, calculating each separately. When calculating the two summations, the sum values are first initialized to 0s, and then updated iteratively, using the variable i as the loop variable.

The true integral of $f(x) = x^2$ has a closed-form representation, $\frac{1}{3}x^3$. Hence the value $\int_0^{10} f(x)dx$ is $\frac{1}{3}10^3 - \frac{1}{3}0^3 = \frac{1}{3}10^3$. An execution of *simpson.py* approximates this value to a high precision.

```
Zhangs-MacBook-Pro:code yue_zhang$ python simpson.py
The integral is equal to 333.333333333
```

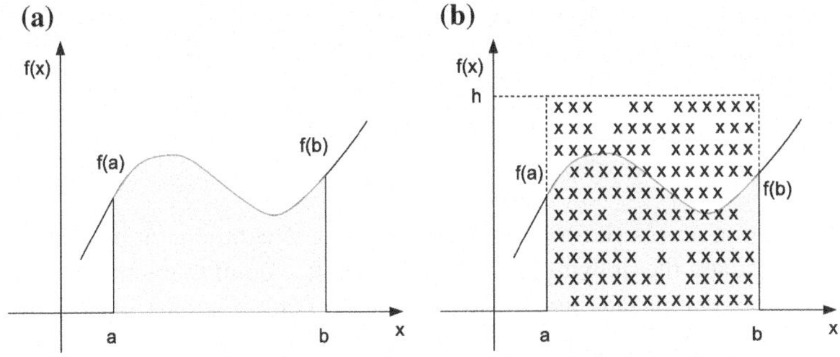

Fig. 5.3 Monte-Carlo integration. **a** The integration. **b** Monte-Carlo integration

5.2.3 Monte-Carlo Methods

A very different method to perform numerical integration is Monte-Carlo integration. **Monte-Carlo methods** are a branch of numerical methods based on random experiments. They have been widely used for the estimation of numerical values, and in particular numerical integration. To give a simple example of estimation by doing experiments, suppose that one wants to estimate whether a coin is fair. She does the estimation by tossing the coin 100 times, and observing the outcomes. If each side of the coin faces up about 50 times, she would deem that the coin is fair.

By doing this experiment, the underlying mathematical quantity that is estimated is the probability of the coin facing up when being tossed. It can be proved that the empirical frequency in the experiments approximates the underlying probability, and the more samples are drawn from the experiments, the less influence random noise can have, and hence the more accurate the estimation is.

Computers can be used to collect a large number of samples automatically, and hence Monte-Carlo methods are frequently used by programs for numerical analysis. For numerical integration, the mathematical quantity to estimate is the area under a function over of span of input. As illustrated by Fig. 5.3, the task of estimating $\int_a^b f(x)dx$ is the task of estimation the area under $f(x)$ over the interval between a and b (i.e. the shaded area in Fig. 5.3a). The area can be estimated by drawing a large number of sample points in a box that contains this area, as shown by Fig. 5.3b. If a sufficiently large number of sample points is drawn, the ratio between the number of sample points that fall inside the shaded area and the total number of sample points approximates the ratio between the area under the function and the area of the containing box.

Denoting the height of the box as h, the total number of samples as N, and the number of samples that fall inside the shaded area as M, the observation above can be formulated as:

$$\frac{M}{N} \approx \frac{\int_a^b f(x)dx}{(b-a) \cdot h},$$

from which an estimation of $\int_a^b f(x)dx$ can be obtained with:

$$\int_a^b f(x)dx \approx h(b-a)\frac{M}{N}$$

This equation forms the basis of Monte-Carlo integration. A program can be written to simulate the drawing of N sample points, counting the number of times M that a sample point falls inside the area under $f(x)$. The basic task behind this program is counting $\sum_{i=0}^{N-1}$ *sample-in-area*(i), and the technique of counting has been discussed in the previous section. However, two detailed questions need to be answered before a full program can be written. First, how can a sample point be drawn? Second, how can one decide whether a sample point is inside the shaded area?

To answer the first question above, one sample point is one random point (x, y), where both x and y are random numbers, with $a \le x \le b$ and $0 \le y \le h$. x and y can be drawn separately using the *random* module. In particular, the *random.random()* function gives a random floating point number between 0 and 1. To scale the number to a range between a and b, the linear transformation function $x' = (b-a)x + a$ can be used. The drawing of the random number y can be performed by using the same technique.

To answer the second question above, a point (x, y) is in the shaded area, or under the function f, if and only if $f(x) > y$. Based on these observations, a program can be written to calculate $\int_0^{10} x^2 dx$ by using the Monte-Carlo method as follows.

```
[mcintegrate.py]
import random
N=10000
a=0
b=10
h=150
M=0
i=0
while i<N:
    # draw one sample
    x=random.random() * (b-a) + a
    y = random.random() * (h-0) + 0
    # count
    if y < x*x:
        M += 1
    i += 1
print 'The integral is', float(h)*(b-a)*M/N
```

mcintegrate.py is based on a loop that repeats for N iterations. Before the iterations begin, the total count M is initialized to 0. In each iteration, it draws a random sample point according to the method described above. If the sample point is under the function $f(x) = x^2$, 1 is added to the count M. Five executions of the program can yield the following results.

```
Zhangs-MacBook-Pro:code yue_zhang$ python mcintegrate.py
The integral is 321.75
Zhangs-MacBook-Pro:code yue_zhang$ python mcintegrate.py
The integral is 333.3
Zhangs-MacBook-Pro:code yue_zhang$ python mcintegrate.py
The integral is 338.4
Zhangs-MacBook-Pro:code yue_zhang$ python mcintegrate.py
The integral is 336.9
Zhangs-MacBook-Pro:code yue_zhang$ python mcintegrate.py
The integral is 322.95
```

Because the Monte-Carlo method is based on random experiments, the outcome of each execution can be different. As shown earlier in this section, the true value of the integral $\int_0^{10} x^2 dx$ is 1000/3, and the estimations approximate this value. When the number of samples increases from 10^4 to 10^6, the estimation becomes more accurate, and five runs of *mcintegrate.py* can yield the following results.

```
Zhangs-MacBook-Pro:code yue_zhang$ python mcintegrate.py
The integral is 334.023
Zhangs-MacBook-Pro:code yue_zhang$ python mcintegrate.py
The integral is 332.8845
Zhangs-MacBook-Pro:code yue_zhang$ python mcintegrate.py
The integral is 333.786
Zhangs-MacBook-Pro:code yue_zhang$ python mcintegrate.py
The integral is 333.7845
Zhangs-MacBook-Pro:code yue_zhang$ python mcintegrate.py
The integral is 332.922
```

Another example of using Monte-Carlo methods is the estimation of π, a mathematical problem that has been investigated over hundreds of years. Again, a Monte-Carlo solution estimates the value indirectly by drawing a large number of random points and calculating the ratio between two shapes.

As shown by Fig. 5.4, the area of the unit square is 1, and the area of the shaded quarter-circle inside the unit square is $\frac{1}{4}\pi 1^2 = \frac{\pi}{4}$. Suppose that N random points are drawn uniformly in the square, of which M fall inside the shaded area. The ratio

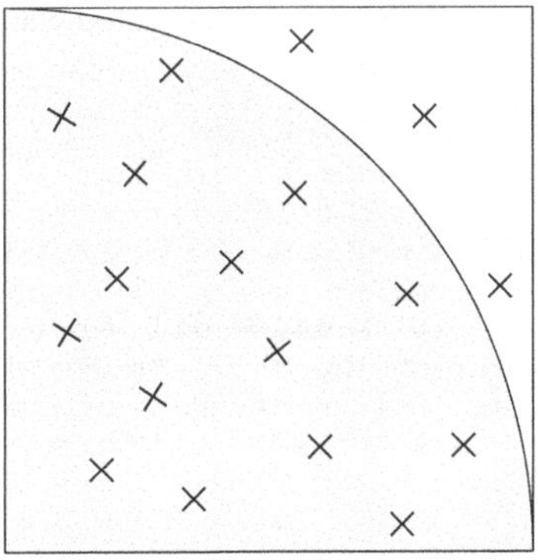

Fig. 5.4 Monte-Carlo estimation of π

between M and N is also the ratio between the area of the shaded quarter-circle and the area of the square.

$$\frac{M}{N} \approx \frac{\pi}{4}/1$$

As a result, the value of π can be estimated by:

$$\pi \approx 4\frac{M}{N}$$

A Python program that estimates the value of π based on the observation above can be written as:

```
[mcpi.py]
import random
import math
N=10000
M=0
i=0
while i<N:
    x = random.random()
    y = random.random()
    if x*x + y*y < 1:
        M += 1
    i += 1
pi = 4 * float(M) / N
print 'The value of pi is approximately %f, which is %
    f different from the precise value' % (pi, math.pi
    - pi)
```

mcpi.py is based on a loop that repeats for N iterations. It uses the variable i as the loop variable, and the variable M to record the number of sample points that fall inside the shaded area. At each iteration, a random point is drawn by drawing the x and y coordinates separately. Each coordinate is drawn between 0 and 1, and the function *random.random()* can be used directly for this purpose. To decide whether a random point is inside a unit circle, the Boolean condition $x^2 + y^2 < 1$ is used. *mcpi.py* reports the estimated value of π, and also the numerical error using *math.pi*. Five executions of *mcpi.py* can yield the following results:

```
Zhangs-MacBook-Pro:code yue_zhang$ python mcpi.py
The value of pi is approximately 3.147600, which is
    -0.006007 different from the precise value
Zhangs-MacBook-Pro:code yue_zhang$ python mcpi.py
The value of pi is approximately 3.162000, which is
    -0.020407 different from the precise value
Zhangs-MacBook-Pro:code yue_zhang$ python mcpi.py
The value of pi is approximately 3.148400, which is
    -0.006807 different from the precise value
Zhangs-MacBook-Pro:code yue_zhang$ python mcpi.py
The value of pi is approximately 3.147200, which is
    -0.005607 different from the precise value
Zhangs-MacBook-Pro:code yue_zhang$ python mcpi.py
```

```
The value of pi is approximately 3.137600, which is
    0.003993 different from the precise value
```

Increasing the number of sample points from 10^4 to 10^6 gives better approximations, and five consecutive runs of *mcpi.py* with $N = 1000000$ can yield the following results:

```
The value of pi is approximately 3.141664, which is
    -0.000071 different from the precise value
Zhangs-MacBook-Pro:code yue_zhang$ python mcpi.py
The value of pi is approximately 3.141672, which is
    -0.000079 different from the precise value
Zhangs-MacBook-Pro:code yue_zhang$ python mcpi.py
The value of pi is approximately 3.142908, which is
    -0.001315 different from the precise value
Zhangs-MacBook-Pro:code yue_zhang$ python mcpi.py
The value of pi is approximately 3.140348, which is
    0.001245 different from the precise value
Zhangs-MacBook-Pro:code yue_zhang$ python mcpi.py
The value of pi is approximately 3.143136, which is
    -0.001543 different from the precise value
```

5.2.4 Differential Equations and Iterative Root Finding

Differential equations are a common tool in scientific modeling. They address a wide range of problems, for which the values of a function at specific points, together with the rate of change of the function, are known, but the function itself is unknown. The general form of differential equations can be written as:

$$f'(x) = g(x)$$
$$f(0) = y_0$$

It can be derived from the definition above that $f(t) = f(0) + \int_0^t g(x)dx$. For some functions $g(x)$, such as $g(x) = x^2$, $\int g(x)dx$ is known, and hence $f(x)$ has an exact solution. However, in practice $\int g(x)dx$ often has no closed form. As a result, numerical integration using the approaches introduced earlier in this section is one common solution.

For small values of t, an alternative solution to find $f(t)$ is to use an iterative method to approximate $f(t)$ directly. As shown in Fig. 5.5, the basic idea is to divide the range $[0, t]$ into n equal intervals, and iteratively approximate the value of $f(x)$ at the end of each interval. Denote the sequence of corresponding x values between 0 and t as $x_1, x_2, \ldots, x_n$, with $x_0 = 0$ and $x_n = t$, the task is to estimate the value of $f(x_n)$ by iteratively estimating the sequence of values $f(x_1), f(x_2), \ldots, f(x_n)$. The differential equation $f'(x) = g(x)$ gives a difference equation for estimating $f(x_i)(i \in [1, n])$:

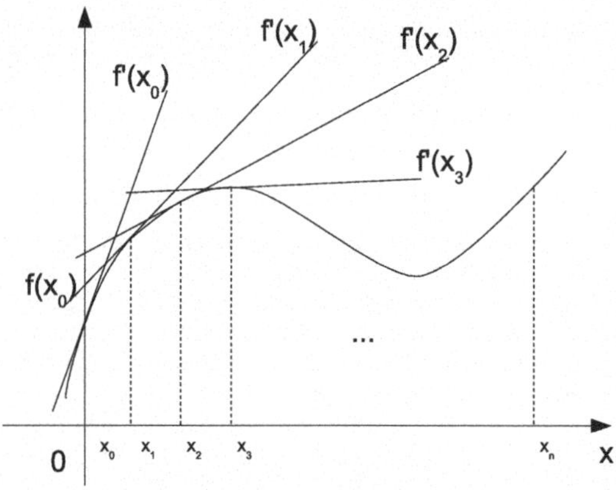

Fig. 5.5 Euler method for differential equations

$$f(x_i) \approx f(x_{i-1}) + dg(x_{i-1}),$$

where $d = t/n$. This approximation is known as *Euler's method.* Given the different equation, the iterative method for calculating number sequences introduced earlier in this chapter can be applied to solve the estimation problem of $f(x_i)(i \in [1, n])$. Suppose that $g(x) = x^2$ and $f(0) = 1$, the following program calculates the value of $f(0.5)$:

```
[euler.py]
f0=1
t=0.5
n=10000
d=t/n
i=0
xi=0
fi=f0
while i<n:
    fi = fi+d*xi*xi
    xi+=d
    i+=1
print 'The value of f(%.2f) is %.5f' % (t, fi)
```

euler.py runs for 10000 iterations, using the variable xi to represent the value x_i and the variable fi to represent the value $f(x_i)$ at each iteration. The two variables are updated iteratively, with $x_i \leftarrow x_{i-1} + d$ and $f(x_i) \leftarrow f(x_{i-1}) + dg(x_{i-1})$. An execution of *euler.py* gives the following result:

```
Zhangs-MacBook-Pro:code yue_zhang$ python euler.py
The value of f(0.50) is 1.04166
```

Because $\int x^2dx$ has a known form $\frac{1}{3}x^3$, the value of $f(0.5)$ is $1 + \frac{1}{3} \times 0.5^3 = 1.0416666666666667$. The estimation using *euler.py* gives a highly accurate answer.

Another useful application of the same iterative approach is *Newton's method* for **finding the root** of a function. Suppose that the function is $f(x)$ and its derivative is known as $f'(x)$. Given an initial guess of a root x_0, the method iteratively finds better approximations to the real root value:

$$x_i \leftarrow x_{i-1} - \frac{f(x_{i-1})}{f'(x_{i-1})}$$

If the initial guess x_0 is sufficiently close to a root of $f(x)$, Newton's method is guaranteed to converge to the real root.

newton.py calculates a root of $f(x) = x^2 - 5x + 6$ given the initial guess $x_0 = 1.23$.

```
x0=1.23
i=0
xi=x0
while True:
    fi = xi*xi-5*xi+6   #given by $f(x)$
    dfi=2*xi-5  #calculated manually from $f(x)$
    print 'Iteration %d, x=%.5f, f(x)=%.12f' % (i, xi, fi)
    xi = xi - fi/dfi  #the difference equation
    i += 1
    if abs(fi-0)<1E-12:
        break
```

newton.py uses the variable xi to represent the values x_i ($i \geq 0$) dynamically, the variable *fi* to represent the values of $f(x_i)$, and the variable *dfi* to represent the values of $f'(x_i)$. Starting from $x_0 = 1.23$, *newton.py* iteratively updates the values of *xi*, *fi* and *dfi*, until *fi* is sufficiently close to 0 (*fi* $\leq 10^{-12}$). Due to rounding off errors, it is unlikely to obtain an *fi* value of 0 exactly. Hence the criteria $y \leq 10^z$, where z is a very small number, is commonly used in numerical analysis to decide whether the number y tends to 0.

newton.py prints the values of *xi* and *fi* at each iteration. Note that a *break* statement under the loop-terminating condition is used to stop the loop, rather than using a Boolean condition directly in the *while* line. This is a convenient way of writing a loop in Python when the loop-terminating condition should be checked at the end of every loop instead of at the beginning. An execution of *newton.py* gives the following result.

```
Zhangs-MacBook-Pro:code yue_zhang$ python newton.py
Iteration 0, x=1.23000, f(x)=1.362900000000
Iteration 1, x=1.76657, f(x)=0.287912519375
Iteration 2, x=1.96285, f(x)=0.038525603991
Iteration 3, x=1.99872, f(x)=0.001286040253
Iteration 4, x=2.00000, f(x)=0.000001645435
Iteration 5, x=2.00000, f(x)=0.000000000003
Iteration 6, x=2.00000, f(x)=0.000000000000
```

5.3 Tuples and the *for* loop

5.3.1 Tuples

Six types of objects have been introduced so far in this book, including the number types *int*, *float*, *long*, and *complex*, the *string* type for text IO and the *Boolean* type for control flow. **Tuples** are another type of objects in Python, which represents an ordered **container** of objects. In particular, a *tuple* object represents a sequence of objects, which can be heterogeneous. Besides *tuples*, Python provides a number of other container types, which will be introduced later in this book.

A **tuple literal** is written as a comma-separated list of literals or identifiers, enclosed in a pair of round brackets:

```
>>> t=(1,2,3)
>>> t
(1, 2, 3)
>>> type(t)
<type 'tuple'>
>>> x='a'
>>> y=True
>>> u = (1.5, x, y, 1+2j)
>>> u
(1.5, 'a', True, (1+2j))
>>> type(u)
<type 'tuple'>
>>> v=()
>>> v
()
>>> type(v)
<type 'tuple'>
>>> w=(1.3,)
>>> w
(1.3,)
>>> type(w)
<type 'tuple'>
>>> d=(1.3)
>>> d
1.3
>>> type(d)
<type 'float'>
```

In the example above, *t* is a tuple that consists of three items, 1, 2, 3, which are all integer objects. *u* is a tuple that consists of four items, 1.5, 'a', True and 1+2j, which belong to different types. The literal of *u* consists of a floating point number literal (1.5), an identifier (x) that represents a string ('a'), an identifier *y* that represents a Boolean object (True) and a complex number literal ($1 + 2j$). When an identifier is used in a tuple literal, the object that the identifier is bound to is taken as the corresponding item in the tuple.

v is en empty *tuple*—a *tuple* that contains no items. w is a *tuple* that consists of a single item (1.3). When writing a *tuple* literal with a single item, a comma must be added after the literal or identifier of the item. This is to differentiate the literal

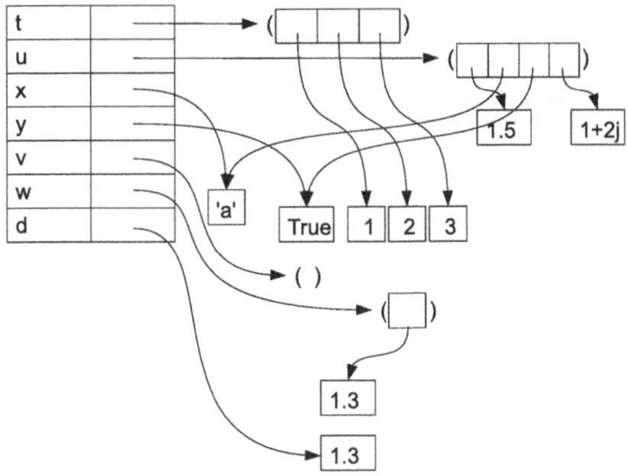

Fig. 5.6 Tuple objects in memory

of a single-item *tuple* from the use of brackets in an expression, which specifies the order of evaluation (c.f. Chap. 2). The value of d, for example, is a floating point number 1.3, because the right hand side of the assignment statement for d consists of an expression with a single floating point literal, enclosed in a pair of brackets.

The memory structure for the example above is shown in Fig. 5.6. Each *tuple* consists of a sequence of object references. When an identifier is used to specify a *tuple* item, the real object that the identifier is bound to is added to the *tuple*. Hence there is no direct link between the identifiers u and x, for example, except that an item of t and x are bound to the same object 'a'.

Tuple operators. *Tuples* support exactly the same set of operations as strings do. In fact, since a string object represents a sequence of characters, a *tuple* object can be regarded as a 'special string object', which contains arbitrary objects rather than characters. For example, the + and * operators apply to *tuples*:

```
>>> t1=(1,2,3)
>>> t2 = (4,5,6)
>>> t3 = t1+t2
>>> t3
(1, 2, 3, 4, 5, 6)
>>> t4 = t1*3
>>> t4
(1, 2, 3, 1, 2, 3, 1, 2, 3)
```

Similar to the case of strings, the + operator concatenates two *tuple* objects into a new *tuple* object, while the * operator makes copies of a *tuple* object. The memory structure of the example above is shown in Fig. 5.7.

The *getitem* and *getslice* operations, and the *len* function introduced in Chap. 3, also apply to *tuples*.

```
>>> t = ('a', 0, 1+3j)
>>> t[0]
'a'
>>> t[1:3]
(0, (1+3j))
>>> t[-2:]
```

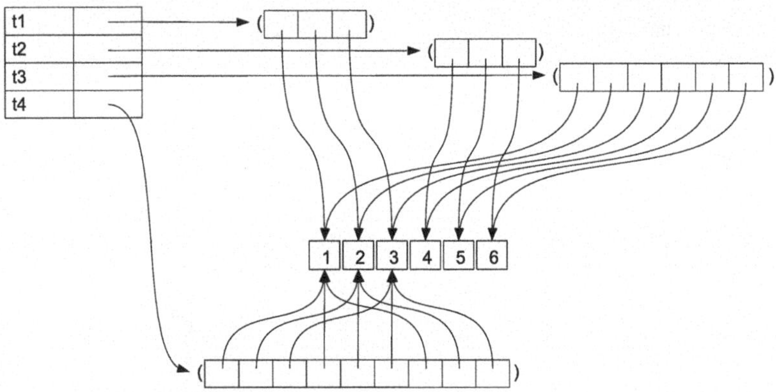

Fig. 5.7 Addition and multiplication for *tuple* objects

```
(0, (1+3j))
>>> len(t)
3
>>> len(t+t)
6
>>> len(t*3)
9
>>> len(())
0
>>> len((1,))
1
```

The *in* and *not in* operators introduced in the previous chapter can also be applied to *tuples*, indicating whether an object is an item in a *tuple* or not.

```
>>> t=(1,'a',True)
>>> 1 in t
True
>>> 1.0 not in t
False
>>> 'a' in t
True
>>> 'ab' in t
False
>>>
```

Conversion between tuples and other types. A *tuple* object cannot be converted to a number. However, it can be converted into a string or a Boolean type object. When a *tuple* is converted into a string, its literal form is taken, with each element in its own literal form. When a tuple is converted into a Boolean object, the result is *False* only if the tuple is empty, and *True* otherwise. Such conversions offer a handy way of making use of implicit conversions, such as putting *tuples* in *print* and *while* statements where string and Boolean objects are expected.

```
>>> x='a'
>>> y=True
```

```
>>> str((x,y,-0.3))
"('a', True, -0.3)"
>>> bool((1,2,3))
True
>>> bool(())
False
>>> print (x,y,-0.3)
('a', True, -0.3)
>>> if (1,2,3):
        print True
True
```

While numbers and Boolean objects cannot be converted to *tuples*, strings can be converted to *tuples*, with each character being mapped into one element in the resulting *tuple* object.

```
>>> tuple(1)
Traceback (most recent call last):
  File "<stdin>", line 1, in <module>
TypeError: 'int' object is not iterable
>>> tuple(True)
Traceback (most recent call last):
  File "<stdin>", line 1, in <module>
TypeError: 'bool' object is not iterable
>>> tuple('abc')
('a', 'b', 'c')
```

When the assignment statement was discussed in Chap. 3, it was mentioned that the left hand side of an assignment statement can be an identifier or a *tuple*. In the latter case, the left hand side must be a *tuple* of identifiers, and the right hand side must be a *tuple* of the same size as the left hand side. Each object in the right-hand-side *tuple* is assigned to the corresponding identifier in the left-hand-side *tuple*. for example,

```
>>> (x, y)=(1, 2)
>>> x
1
>>> y
2
>>> (x, y, z) = (3, 'a', x)
>>> x
3
>>> y
'a'
>>> z
1
```

When a *tuple*-to-*tuple* assignment is executed, Python first tries to find the objects that are contained in the right-hand-side *tuple*. In the second example above, it finds the *int* object 3, and string object 'a' and the *int* object 1, as referred to by the identifier x. Then Python assigns the three objects to the corresponding identifiers in the left-hand-side *tuple*, resulting in x being 3, y being 'a' and z being 1. Note that the value of x changes in the execution of this assignment statement, in a similar way to the execution of the statement $x = x + 1$.

Tuple assignment provides a convenient way of swapping two variables.

```
>>> x=1
>>> y=2
>>> (x,y)=(y,x)
>>> x
2
>>> y
1
```

Using the trick, the induction of Fibonacci numbers introduced earlier in this chapter can be written without a temporary variable. In particular, the three statements

```
t=xi
xi=xi+xii
xii=t
```

can now be replaced with one statement:

```
(xi, xii) = (xi+xii, xi)
```

In a *tuple*-to-*tuple* assignment, the left-hand-side *tuple* can be written without brackets. For example, the statement above can also be written as:

```
xi, xii = (xi+xii, xi)
```

5.3.2 *The* for *Loop*

The for statement is a second looping statement introduced in this book. It can be used to iterate through a container object. The syntax of the *for* statement is shown in Fig. 5.8, which begins with a line that consists of the *for* keyword, followed by an identifier, and then the *in* keyword, and then a container object, and finally a colon. After the *for* line is an indented statement block, which is executed repeatedly, once for each element in the collection object. In each iteration, the current element of the collection is assigned to the identifier after the *for* keyword, which serves as the loop variable. For example, the following code iterates through a *tuple*, printing out each element:

```
>>> t=(1,'a',True)
>>> for x in t:
...     print x
...
1
a
True
```

In the example above, the *for* statement consists of two lines, the first being the *for* line with the loop variable being defined as x, and the second line being an indented statement block that consists of a single statement. When executed, the *for* statement iterates through all elements of t, assigning the current element to the identifier x at each iteration, before executing the indented statement block, or *loop body*. As a result, the dynamic execution sequence of the example above is:

```
x=1
print x
x='a'
print x
x=True
print x
```

An equivalent loop using the *while* statement is:

```
>>> t=(1,2,3)
>>> i=0
>>> while i<len(t):
...     x=t[i]
...     print x
...     i += 1
...
1
2
3
```

In the example above, the variable i is used as the loop variable. It iterates through the values 0, 1, ..., $len(t)$. At each iteration, a corresponding element $t[i]$ is taken from the container t, and assigned to the variable x. This type of iteration through a container object is called *traversal*. Compared with the *while* loop, the *for* loop is more convenient for performing *tuple* traversal, since using a *for* statement, the additional loop variable x is unnecessary. Container objects that can be iterated through using a *for* loop are also called **iterable objects**. *Tuples* are iterable.

The *break* and *continue* statements can also be applied to a *for* loop, with the same interpretation as in a *while* loop: the *break* statement terminates the loop, and the *continue* statement skips the rest of the loop body and starts the next loop iteration.

```
>>> t=(1,2,3,4,5,7,10)
>>> for i in t:
...     if i>5:
...         break
...     print i
...
1
2
3
4
5
>>> for i in t:
...     if i
```

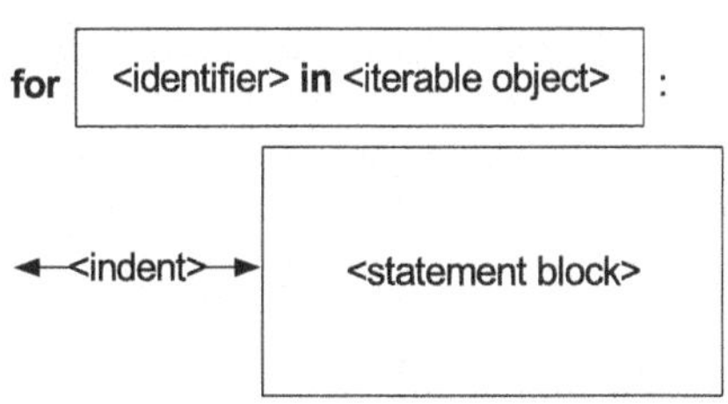

Fig. 5.8 The *for* statement

```
...         continue
...     print i
...
2
4
10
```

The example above contains two *for* loops over the *tuple* t. In the first loop, when the first element in t that is greater than 5 is enumerated, the loop terminates. In the second loop, whenever the current element is odd, the loop continues without it being printed.

Similar to the *while* loop, an *else* line, followed by an indented statement block, can also be added to the end of a *for* loop. If the *for* loop finishes enumerating all the elements in the container object (i.e. no break happens during the loop execution), the indented statement block under the *else* line is executed.

```
>>> t=(1,2,3)
>>> for x in t:
...     print x
... else:
...     print 'done'
...
1
2
3
done
>>> for x in t:
...     print x
...     break
... else:
...     print 'done'
...
1
```

The example above consists of two *for* statements. When the first is executed, the loop finishes and hence the *else* statement block is executed. When the second *for* loop is executed, an unconditional *break* statement is executed in the first loop iteration, and hence the loop finishes immediately, without the *else* statement block being executed.

5.3.3 Problem Solving by Traversal of a Tuple

The *for* loop can be used to solve all the basic problems introduced earlier in this chapter, including summation, production, maximum finding and search, all by traversal of a container object. The following example illustrates the use of a *for* loop for summing up all elements in a tuple.

```
>>> t=(1.5,2.0,3.1,6.4,-4.7,7.1,-0.01)
>>> s=0
>>> for x in t:
```

```
...         s += x
...
>>> print s
15.39
```

The *for* statement above iterates through t, assigning to x the value of the current element being enumerated at each iteration. Summation is performed by accumulating all x values to the sum s, which is initialized to 0. The thinking behind the method is the same as the summation using a *while* loop introduced earlier in this chapter.

For summation, Python provides a built-in function *sum*, which takes a single iterable argument and returns the sum of all elements in the container object. Using the *sum* function, the example above can be written as:

```
>>> t=(1.5,2.0,3.1,6.4,-4.7,7.1,-0.01)
>>> sum(t)
15.39
```

In similar spirit, Python provides two built-in functions *max* and *min*, which take a single iterable argument and return the maximum and minimum elements in the container object, respectively. As introduced earlier in this chapter, the *max* and *min* functions can also take multiple numerical arguments and return their maximum/minimum.

The following code illustrates the use of a *for* loop to find the maximum value of a tuple. The thinking behind is the same as finding a maximum value using a *while* loop.

```
>>> t=(1.5,2.0,3.1,6.4,-4.7,7.1,-0.01)
>>> s=float('-inf')
>>> for x in t:
...     if x > s:
...         s = x
...
>>> print s
7.1
```

The problem above can also be solved using the function call $max(t)$.

The following code illustrates the use of a *for* loop to decide whether there is a multiple of 7 in a *tuple* of integers. The thinking behind is the same as searching for a target using a *while* loop.

```
>>> t=(1,75,31,614,59,63,110)
>>> for x in t:
...     if x % 7 == 0:
...         print 'Yes'
...         break
... else:
...     print 'No'
...
Yes
```

Two built-in functions, *all* and *any*, take a single container argument and return *True* if and only if all/any element in the container is *True*. Non-Boolean values are converted to the Boolean type automatically. The problem above can be solved by the use of the *any* function, and the *map* function introduced in Chap. 9, where *functional programming* concepts are introduced.

Exercises

1. Calculate
   (a) the sum of all odd numbers between 0 and 100 that make $sin(x) > 0$;
   (b) the product of the list (1, 2.5, -3.3, -4, 5, 6, 10);
   (c) $\frac{d}{dx} log(x^3 - 2x + 5) \cdot sin(x)$ at $x = 0.5$;
   (d) $\int_1^{100} log(x)dx$ numerically.
2. Estimate the following values using summation. What is the correlation between n and the error?
   (a) $\pi \approx 4 \cdot \sum_{i=1}^{n} \frac{(-1)^{i+1}}{2^i - 1}$
   (b) $\pi \approx 2\sqrt{3} \sum_{i=0}^{n} \frac{(-1)^i}{3^i(2i+1)}$
   (c) $sin(x) \approx \sum_{i=0}^{n}((-1)^i x^{(1+2i)})/((1+2i)!)$
   (d) $sin(x) \approx \sum_{i=0}^{n}((-1)^i(-pi/2+x)^{(2i)})/((2i)!)$
   (e) $log(1+x) \approx \sum_{i=1}^{n}(-1)^{i-1}\frac{x^i}{i}$
   (f) $exp(x) \approx \sum_{i=0}^{n}\frac{x^i}{i!}$
3. What are the values of i after the following programs are executed?
   (a) print 1, 3, 5, 7, 9

```
t = (1,3,5,7,9)
for i in t:
    print i
print i
```

   (b) still print 1, 3, 5, 7, 9

```
t = (1,3,5,7,9)
for i in t:
    print i
    i += 1
print i
```

   (c) print 2, 4, 6, 8, 10

```
t = (1, 3, 5, 7, 9)
for i in t:
    i += 1
    print i
print i
```

   (d) print 1, 3, 5, 7, 9

```
t = (1, 3, 5, 7, 9)
for i in t:
    print i
else:
```

```
    i += 1
print i
```

(e) print 1, 3, 5

```
t = (1, 3, 5, 7, 9)
for i in t:
    if i > 5:
        break
    print i
print i
```

(f) still print 1, 3, 5

```
t = (1, 3, 5, 7, 9)
for i in t:
    if i > 5:
        continue
    print i
print i
```

4. Write a program that
   (a) asks the user for an integer, and then displays the sum of all the digits;
   (b) asks the user for two integers, and then displays whether the second integer is a divisor of the first integer;
   (c) asks the user for an integer, and then displays its largest divisor (Hint: consider solution of (b));
   (d) asks the user for an integer, and then displays whether the integer is a prime number (Hint: consider solution of (c));
   (e) repeatedly asks the user to enter an integer, until the user enter 0, after which the maximum of all the numbers is shown;
   (f) asks the user for two integers, and then displays the greatest common divisor of the two integers (Hint: consider the solution of (c)).
5. Write a program that draws the following multiplication table:

```
    1  2  3  4  5  6  7  8  9
 ****************************
1*  1  2  3  4  5  6  7  8  9 *
2*  2  4  6  8 10 12 14 16 18 *
3*  3  6  9 12 15 18 21 24 27 *
4*  4  8 12 16 20 24 28 32 36 *
5*  5 10 15 20 25 30 35 40 45 *
6*  6 12 18 24 30 36 42 48 54 *
7*  7 14 21 28 35 42 49 56 63 *
8*  8 16 24 32 40 48 56 64 72 *
9*  9 18 27 36 45 54 63 72 81 *
 ****************************
```

6. Write a program that asks the user for an integer, and then displays its binary value in 2's complement form, as introduced in Chap. 2, assuming that the integer can be represented using 32 bits. Hexadecimal numbers are a more succinct notation. Each digit in a hexadecimal number ranges from 0-15. The value 10, 11, 12, 13, 14

and 15 are denoted with 'A', 'B', 'C', 'D', 'E' and 'F', respectively. For example, a hexadecimal value 1B is equal to $1 \times 16 + 11 \times 1 = 27$, and a hexadecimal value 2A1 is $2 \times 16^2 + 10 \times 16 + 1 = 417$. Extend the program above so that it displays the hexadecimal form of the input number. Is it easier to convert a decimal number into a hexadecimal number, or to convert a binary number into a hexadecimal number?

Chapter 6
Functions

As introduced in Chap. 2, function calls enrich the power of Python expressions by introducing more functionalities. The mathematical functions provided by the *math* module, such as *exp*, *log* and *sin*, allow a Python expression to perform arithmetic calculation beyond the power of mathematical operators. In addition, built-in functions such as *len*, *id*, *int* and *raw_input*, provide mechanisms to get information about Python objects, perform type conversion, and interact with users. This chapter introduces the ways to define custom functions in Python, allowing customized functionalities to be modularized and reused.

6.1 Function Definition Using *lambda* expressions

To begin with, consider the problem of defining $f(x, y) = x+2y$. The function takes two number arguments, and returns their weighted sum. The name of the function is f. Python provides a simple type of expression, the **lambda expression**, for the definition of such a function. The syntax of the *lambda* expression is:

```
lambda <arguments>: <return value expression>
```

A *lambda* expression begins with the keyword *lambda*, followed by a comma-separated list of **parameters**, and then a colon, and finally an expression that specifies the return value for calls to the function. The value of a *lambda* expression is a **function object**. For example, to specify the function *f* above, the following assignment statement can be used:

```
f = lambda x, y: x+2*y
```

In the example above, the right hand side of the assignment statement is a *lambda* expression that defines a function object, and the left hand side is an identifier *f*. The value of the function object is assigned to the identifier. After the function definition and the assignment are executed, *f* can be used in function call expressions.

Y. Zhang, *An Introduction to Python and Computer Programming*,
Lecture Notes in Electrical Engineering 353, DOI 10.1007/978-981-287-609-6_6

```
>>> f = lambda x, y: x+2*y
>>> f(1, 2)
5
>>> f(3, 4)
11
```

The mechanism of function calls. In the example above, the function object *f* is called twice. Each time a function call is invoked, Python passes the input arguments to the function object and evaluates the return value. In the first function call expression $f(1, 2)$, the two arguments 1 and 2 are passed into the function, and assigned to the parameters x and y, respectively. The mechanism of argument passing is the same as the assignment statement: the assignment of the argument values 1 and 2 to the parameters x and y is effectively the same as performing $x = 1$ and $y = 2$ inside the function, respectively.

After argument passing, Python evaluates the expression $x + 2 * y$ to determine the return value of this function call, obtaining the integer object 5. Thus the function call finishes with the value of the expression $f(1, 2)$ being 5, which is shown by IDLE. The same process happens to the second function call $f(3, 4)$, with 3 and 4 being assigned to the parameters x and y inside the function, respectively, leading to a different return value by the evaluation of $x + 2 * y$. It can be seen from this example that each time a function call expression is evaluated, the return value expression is evaluated afresh, with new argument values being assigned to the parameters.

Note the difference between a *function definition* and a *function call*. The former specifies a function object, while the latter invokes a function in order to obtain a return value. The return value expression in a *lambda* expression typically includes all the parameters, which represents the input arguments, for the intuitive reason that a function represents certain operations on the arguments. However, the return value of a function can also be independent of its arguments. For example, the following function returns the constant value 2.

```
>>> f = lambda x:2
>>> f(1)
2
>>> f(2)
2
```

The example above is not practically useful, but illustrates the correlation between input arguments and the return value expression. Another example where the input arguments do not affect the return value is the built-in function *raw_input*, which displays the input argument as a prompt to the user, but returns user input.

The number of input arguments in a function call must be the same as the number of parameters specified in the function definition. Otherwise, Python will report an error in the function call.

```
>>> g=lambda x,y:x
>>> g(1,2)
1
>>> g(1)        # too few arguments
Traceback (most recent call last):
  File "<stdin>", line 1, in <module>
```

```
TypeError: <lambda>() takes exactly 2 arguments
   (1 given)
>>> g(1,2,3)   # too many arguments
Traceback (most recent call last):
  File "<stdin>", line 1, in <module>
TypeError: <lambda>() takes exactly 2 arguments
   (3 given)
```

Function objects. Similar to other objects, a function object defined by a *lambda* expression can be assigned to different identifiers.

```
>>> f = lambda x, y: x+y
>>> g = f
>>> f(1,2)
3
>>> g(1,2)
3
```

In the example above, both f and g refer to the same function object, and calls to $f(1,2)$ and $g(1,2)$ lead to the same result.

The types of objects introduced so far in this book support different operators. For example, string and tuple objects support the *getitem* (i.e. []) operator, the + operator and the * operator. Intuitively, a function objects supports only the function call operator (i.e. ()), but few other operators.

```
>>> f=lambda x, y: x+y
>>> g=lambda x, y: x-y
>>> f+g
Traceback (most recent call last):
  File "<stdin>", line 1, in <module>
TypeError: unsupported operand type(s) for +:
 'function' and 'function'
>>> f*2
Traceback (most recent call last):
  File "<stdin>", line 1, in <module>
TypeError: unsupported operand type(s) for *:
 'function' and 'int'
>>> f[2]
Traceback (most recent call last):
  File "<stdin>", line 1, in <module>
TypeError: 'function' object is not subscriptable
```

Using functions to simplify code. By using function definitions, *simpson.py* in the previous chapter, which calculates the integral of $f(x) = x^2$ between 0 and 10, can be simplified. The original program contains the calculation of four terms, $f(a)$, $2\sum_{i=1}^{n-1} f(a+id)$, $4\sum_{i=0}^{n-1} f(a+(i+\frac{1}{2})d)$ and $f(b)$, whereas $f(x)$ is hard coded as a complex expression in the evaluation of each term. A new version of the program, *simpson_lambda.py*, specifies $sq(x) = x^2$, and uses a function call to sq in the place of hard-coded expressions in the evaluation of all four terms.

```
[simpson_lambda.py]
a=0
b=10
n=10000
d=float(b-a)/n
```

```
sq=lambda x: x*x
#f(a)
fa=sq(a)
#f(b)
fb=sq(b)
# sum f(a+id)
s1=0
i=1
while i<=n-1:
    s1 += sq(a+i*d)
    i += 1
# sum f(a+(i+0.5)d)
s2=0
i=0
while i<=n-1:
    s2 += sq(a+(i+0.5)*d)
    i += 1
# area
s = (d/6) * (fa+2*s1+4*s2+fb)
print 'The integral is equal to', s
```

An execution of *simpson_lambda.py* yields the same result as *simpson.py*.

```
Zhangs-MacBook-Pro:code yue_zhang$ python
    simpson_lambda.py
The integral is equal to 333.333333333
```

simpson_lambda.py allows a much easier modification to the code if the function to be integrated changes from $f(x) = x^2$ to $f(x) = 3x^2 + 2x - 1$. The only change needed is the definition of *sq*.

```
sq=lambda x: 3*x*x+2*x-1
```

The rest of the program stays the same. An execution of the new program yields the following result.

```
Zhangs-MacBook-Pro:code yue_zhang$ python
    simpson_lambda.py
The integral is equal to 1090.0
```

The result is identical to the real integral, 1090, which can be calculated by $g(10) - g(0)$, with $g(x) = x^3 + x^2 - x$ being the integral of $f(x)$. If *simpson.py* is to be modified to find the same integral, the four lines that calculate the values of each term in the calculation of $f(x)$ must be modified significantly.

Default arguments and keyword arguments. As introduced in Chap. 2, some functions allow arguments to be unspecified. For example, the *math.log* function can take two arguments x and y, returning the value $log_y x$.

```
>>> import math
>>> math.log(256, 2)
8.0
```

On the other hand, the second argument can also be omitted in a call to *math.log*, resulting the value $log_e x$.

```
>>> import math
>>> math.log(math.e)
1.0
```

To specify an optional parameter in a function definition, a default argument value can be specified for the parameter, by adding an explicit assignment (i.e. = *value*) to the argument in the argument list. For example:

```
>>> f=lambda x, y=0: x + 2*y
>>> f(1)
1
>>> f(1,2)
5
```

In the example above, the default value of the input argument y is set to 0. As a result, where y is not specified in a function call to f (i.e. $f(1)$), it is assigned to 0 by default.

An argument with a default value is also called a **default argument**. Parameters with default arguments must be put after parameters without default arguments (i.e. without a default value) in a function definition, so that there is no ambiguity when some input arguments are omitted in a function call.

```
>>> f=lambda x,y,z=0,w=0: x+2*y+3*z+4*w
>>> f(1,2)
5
>>> f(1,2,3)
14
>>> f=lambda x,y=0,z=0,w:x+2*y+3*z+4*w
  File "<stdin>", line 1
SyntaxError: non-default argument follows default
    argument
```

In the example above, when both z and w take default values, a function call with three input arguments (i.e. $f(1, 2, 3)$) specifies the value of z (i.e. 3) but not that of w. However, the definition of f becomes illegal if both y and z take default values, but w does not.

A function call expression can specify the assignment of input arguments explicity, using *name=value*, which is useful when there are more than one arguments with default values.

```
>>> f=lambda x,y,z=0,w=0: x+2*y+3*z+4*w
>>> f(1,2,3)
14
>>> f(1,2,w=3)
17
```

In the example above, the third input argument 3 is by default assigned to z. However, if it is assigned to w explicity, as in $f(1, 2, w = 3)$, then z will take the default argument. An argument that is explicitly assigned in a function call is called a **keyword argument**.

6.2 Function Definition Using the *def* statement

As introduced in the previous chapter, many mathematical functions cannot be computed by using a simple expression. There are cases where the value of a custom function must be obtained by iterative numerical computation. In such cases, the calculation of a return value requires a procedure that consists of multiple statements, typically representing a dynamic control flow. The syntax of a *lambda* expression does not support the definition of such a procedure.

Instead, Python provides a special statement, the **def statement**, for the definition of functions that consist of multiple statements. To begin with, consider the following function f, which is equivalent to the *lambda* function *lambda x,y: x+2*y*:

```
>>> def f(x,y):
        z=2*y
        return x+z
>>> f(1,2)
5
>>>f(3,4)
11
```

The syntax of the *def* statement is illustrated in Fig. 6.1. The first line of a *def* statement starts with the *def* keyword, followed by a custom function name, and a comma-separated list of parameters, enclosed in a pair of round brackets, and finally a colon. After the *def* line is an indented statement block, which is called the *function body*. The function body contains one or many **return statements**, which begin with the *return* keyword, followed by an expression:

```
return <expression>
```

The *def* statement performs two tasks. First, it defines a *function object*. Second, it associates the function object with the function name given in the *def* line. In contrast, a *lambda* expression only defines a function object, and an assignment statement is needed to give a name to the function.

In a function call, the function name is used as the identifier for accessing the function object. The parameters defined in the *def* line are assigned the argument values given in the function call expression, in the same way as the argument assignment process of a *lambda* function, before the function body is executed. During the execution of the function body, if one *return* statement is executed, the execution

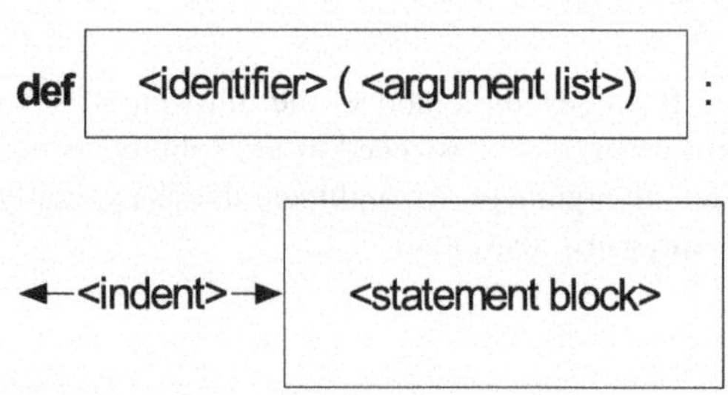

Fig. 6.1 The *def* statement

of the function body is terminated, with the value of the expression in the *return* statement being taken as the return value of the function call.

In the example above, the *def* statement defines a function object with two parameters, and assigns the name *f* to it. When the call *f(1,2)* is evaluated, the argument 1 and 2 are assigned to the parameters *x* and *y*, respectively, and the function body is executed when the *return* statement is executed, the function call returns, with the value of $x + z$, which is 5.

The ***None object*** *and the* ***pass*** *statement.* If no *return* statements are executed before the whole function body is executed, a special return value, *None*, is taken as the return value of the function call. Here *None* belongs to a new type, the *NoneType*. It is the only object that belongs to the type.

```
>>> type(None)
<type 'NoneType'>
>>> a=None
>>> a==1
False
>>> a==0
False
>>> a==''
False
>>> a==False
False
```

The object *None* is equal to neither 0, nor '', nor *False*, because *NoneType* is a different type as compared to a number, the string or the Boolean type. *None* can be converted into a string, 'None', or a Boolean object, *False*, which provides a convenient way of putting the *None* object in *print* and *while* statements.

```
>>> str(None)
'None'
>>> bool(None)
False
>>> a = None
>>> a
>>> type(a)
<type 'NoneType'>
>>> print a
None
```

In the example above, when the value of *None* is given to IDLE (i.e. by typing the identifier *a* directly as an expression in IDLE), nothing is shown on the console. This is different from other objects, for which IDLE display the value. By default, IDLE does not display the *None* value. However, when given to the *print* statement, *None* is first converted into a string, and then shown.

The *None* object cannot be converted to an integer, a floating point number, or a tuple.

```
>>> int(None)
Traceback (most recent call last):
  File "<stdin>", line 1, in <module>
TypeError: int() argument must be a string or a number
    , not 'NoneType'
```

```
>>> float(None)
Traceback (most recent call last):
  File "<stdin>", line 1, in <module>
TypeError: float() argument must be a string or a
    number
>>> tuple(None)
Traceback (most recent call last):
  File "<stdin>", line 1, in <module>
TypeError: 'NoneType' object is not iterable
```

None serves as a special value in Python, representing the meaning 'nothing'. It therefore does not make sense to convert any other types to *NoneType*.

Correlated to the *None* object is a statement that does nothing: the **pass** statement. The syntax of the *pass* statement consists of the *pass* keyword only. It serves as a placeholder where a statement block is required, but nothing needs to be performed.

A simplest function can be one that takes no argument, and performs nothing:

```
>>> def f():
...     pass
...
>>> f()
>>> a=f()
>>> a
>>> type(a)
<type 'NoneType'>
```

In the example above, the function call $f()$ does not perform anything: the function body is a single *pass* statement. In this example, the *pass* statement is necessary, because an indented statement block is required in the definition of a function. Since no *return* statement is executed during the execution of the function body, the return value of the function call $f()$ is *None*. Note that when $f()$ is used as a single-line expression, nothing is shown as the value of the expression. This is because IDLE shows no output for the value *None*, as illustrated earlier.

For a more complex example, consider a function that shows the string 'Hello, world' on the console.

```
>>> def g():
...     print 'Hello, world'
...
>>> a = g()
Hello, world
>>> print a
None
```

In the example above, when the assignment statement $a = g()$ is executed, the expression $g()$ is evaluated first. It is a function call expression, which leads to the invoking of the function body of g. As a result, the string 'Hello, world' is printed to the console. The value of the function call expression $g()$ is assigned to the identifier a. Since no *return* statement is executed, the default return value, *None*, is returned.

6.2.1 The Dynamic Execution Process of Function Calls

Function calls are not sequential control flow because the control flow leaves the main program for the execution of the function body, and then returns to the main program again after the execution of the function being called. The dynamic execution sequence of the code above can be written as:

```
(1) def g():                # statement 1; function definition
(2) print 'Hello, world' # statement 2; evaluating g(),
        executing function body; the return value is None by
        default.
(3) a = None                # statement 2; return value assignment
(4) print a                 # statement 3
```

A function that does not contain any *return* statements is also called a **procedure**: calling of the function performs some tasks, but does not evaluate to a return value. Correspondingly, a call to a procedure is typically used as a single-line expression (e.g. $g()$ in a single line), but not as a part of other expressions or statements. The function g above is a procedure that performs the display of a string message. The built-in function *reload* introduced in Chap. 3 is an other example procedure.

For another example of the dynamic control flow of function calls, consider an expression that contains more than one function calls. In such cases, each function call evaluation invokes the corresponding function body.

```
>>> def f():
...     print 'called'
...     return 1
...
>>> f() + f() + 1
called
called
3
```

When the expression $f() + f() + 1$ in the code above is evaluated, the function $f()$ is called twice. Each time the function body is executed, the message 'called' is displayed. As a result, the message is displayed twice, as shown by the outputs.

A function body can contain calls to other functions, leading to more complex dynamic execution sequences. Nevertheless, the mechanism for each function call evaluation remains the same: the control flow leaves the current statement for an execution of the function being called, and then returns, with the return value as the value of the fuction call expression. For example:

```
>>> def f():
...     print 'Enter f'
...     print 'Leave f'
...
>>> def g():
...     print 'Enter g'
...     f()
...     print 'Leave g'
...
>>> g()
Enter g
```

```
Enter f
Leave f
Leave g
```

The dynamic execution sequence of the code above is illustrated by the messages printed out. In particular, the function call $f()$ is nested as a part of the function call $g()$, because it is invoked in the execution of the function body of g. More discussions on nested function calls are given in the Sect. 6.3.

6.2.2 Input Arguments

As introduced earlier, the passing of arguments to a function defined by the *def* statement is the same as the passing of arguments to a *lambda* function: each parameter in the function is *assigned* the corresponding input argument value in the function call.

```
>>> def h(x, y):
...     print "x =", x, ", y =", y
...
>>> a=2
>>> h(1,a)
x = 1 , y = 2
```

In the example above, when the function call expression $h(1, 2)$ is evaluated, the assignments $x = 1$ and $y = a$ are executed before the function body is executed, leading to the result values $x = 1$ and $y = 2$.

Similar to *lambda* functions, default arguments can also be defined for *def* functions.

```
>>> def sum(a, b, c=0):
...     return a+b+c
...
>>> sum(1,2)
3
>>> sum(1,2,3)
6
```

In the example above, the function *sum* can take two or three input arguments. In the former case, the parameter c is assigned the default argument 0, and the return value $a + b + c$ is equal to the sum of a and b.

Note in the example above that the name of the custom function, *sum*, is the same as a built-in function in Python. The built-in *sum* function, as introduced in the previous chapter, takes one tuple argument and returns the sum of all the elements in the tuple. When defined, the custom *sum* function *overrides* the built-in function.

```
>>> def sum(a, b, c=0):
...     return a+b+c
...
>>> l=(1,2,3)
```

```
>>> sum(1)   # built-in sum overridden
Traceback (most recent call last):
  File "<stdin>", line 1, in <module>
TypeError: sum() takes at least 2 arguments (1 given)
```

As introduced in Chap. 3, the overridden built-in function *sum* can still be accessed to, but explicitly via the *__builtins__* module.

```
(continued from above)
>>> l=(1,2,3,4,5,6,7,8,9,10)
>>> __builtins__.sum(l)
55
```

6.2.3 Return Statements

Return and conditionals. The custom function *sum* above uses a single *return* statement to specify the return value of function calls; the function is identical to the *lambda* expression *lambda* $x, b, c = 0$: $a + b + c$. On the other hand, *return* statements in *def* functions can be used more flexibily in dynamic control flows, achieving much more powerful functionalities. The following example shows how *return* statements can be used in branching control flows.

```
>>> def posneg(x):
...     if x>0:
...         return 'positive'
...     elif x<0:
...         return 'negative'
...     else:
...         return 'zero'
...
>>> posneg(10)
'positive'
>>> posneg(-1)
'negative'
>>> posneg(0)
'zero'
```

In the example above, three *return* statements are written in the function *posneg*, all being embedded in an *if ... elif ... else* statement. There are three branches in this *if* statement. However, depending on the input value x, only one branch is executed at an execution of the statement. As a result, the *return* statement in the corresponding branch will be executed, giving a dynamic return value for the function call expression.

Because the *return* statement terminates the execution of a function immediately, no more statements in the function body will be executed after the *return* statement is executed. The fact can be used to simplify functions. For example, the function *posneg* above can also be written as:

```
>>> def posneg(x):
...     if x>0:
```

```
...         return 'positive'
...     if x<0:
...         return 'negative'
...     return 'zero'
...
>>> posneg(10)
'positive'
>>> posneg(-1)
'negative'
>>> posneg(0)
'zero'
```

In the example above, there are three *return* statements in the function *posneg*. The first two *return* statements are each embedded in an *if* statement, and the last *return* statement is used unconditionally. In a dynamic control flow, if the value of the input argument *x* is positive, the *return* statement in the first *if* statement will be executed, which terminates the function call without the second *if* statement being reached. Similarly, if the value of *x* is negative, the *return* statement in the second *if* statement will be executed, which terminates the function call wihtout the last *return* statement being reached. Consequently, only when the value of *x* is neither positive nor negative will the last *return* statement be executed.

Note, however, the fact that *posneg* can be written in the way above is all because the *return* statement terminates a function call. If the *posneg* function is changed to a procedure, which prints out the messages 'positive', 'negative' and 'zero' rather than returning the stings, the second way to write the function body above will cause errors.

```
>>> def posneg(x): # the first way
...     if x>0:
...         print 'positive'
...     elif x<0:
...         print 'negative'
...     else:
...         print 'zero'
...
>>> posneg(10)
positive
>>> posneg(-1)
negative
>>> posneg(0)
zero
>>>
>>> def posneg(x): # the second way
...     if x>0:
...         print 'positive'
...     if x<0:
...         print 'negative'
...     print 'zero'
...
>>> posneg(10)
positive
zero
>>> posneg(-1)
negative
```

```
zero
>>> posneg(0)
zero
```

The second *posneg* function above will always print the message 'zero', because the last *return* statement is unconditional. The execution of the *print* statements does not terminate the function call.

The difference between *return* and *print* statements in a function may seem confusing, especially when the function is called in IDLE: both prints out the value of an expression. However, a more careful look can reveal that the two statements are essentially irrelevant to each other. When the *posneg* functions with *return* statements are called, IDLE displays the return values as the values of the function call expression *posneg*(); but when the *posneg* functions with the *print* statements are called, the messages are printed when the function body is executed, and the return values of the function calls, which are *None*, are not displayed by IDLE.

Return in loops. The *return* statement can also be used in a looping control flow. For example, consider the problem of finding the first odd number in a tuple of integers.

```
>>> def findodd(t):
...     for i in t:            # for each element
...         if i%2==1:         # if odd
...             return i       # return
...
>>> print findodd((1,2,3))
1
>>> print findodd((2,4,6))
None
```

In the example above, *findodd* takes a single input tuple argument. It iterates through the elements in the tuple, using the loop variable *i*. At each iteration, if the value of the current element is odd, it is returned directly. After the *return* statement is executed, no further statements in the function body is executed, and hence the loop body is interrupted. In case no odd numbers are in the input list, the function body finishes without a *return* statement being executed, and hence *None* is returned.

The function *findodd* above is an example of search in the context of a tuple. The usage of the *return* statement in a loop is suitable only when the *first* answer is sufficient. If the requirement changes, and *all* odd numbers in the input tuple are needed, the function must be modified. Here the thinking of generalized summation discussed in the beginning of the previous chapter can be applied: the return value *s* is now a tuple rather than a number, initialized to an empty tuple, and the incremental step to update *s* is tuple concatenation rather than numerical addition.

```
>>> def findodd(t):
...     s=()
...     for i in t:        # for each element
...         if i%2==1:     # if odd
...             s+=(i,)    # update return value
...     return s
...
>>> findodd((1,2,3))
```

```
(1, 3)
>>> findodd((2,4,6))
()
```

In the example above, the function *findodd* calculates the output tuple by performing generalized summation. The incremental step, $s+ = (i,)$, extends s by concatenating its current value with the single-element tuple $(i,)$. A single *return* statement is used at the end of the function.

Tuple Return Values. The example above returns a tuple. When more than one values are returned by a function, they can be put into a tuple. For example, the function below swaps two input arguments.

```
>>> def swap(x,y):
...     return(y,x)
>>> a=1
>>> b=2
>>> (a,b)=swap(1,2)
>>> a
2
>>> b
1
```

6.2.4 Modularity

The freedom of writing a statement block in a *def* statement allows *modularity* in code design: particular functionalities are put into specific procedures and invoked by function calls, making code easier to understand and maintain. For example, consider *grade.py* in Chap. 4, which calculates the grade of a student given an exam score. Suppose that the program needs to be extended, so that it repeatedly asks the user to enter a score, returning the corresponding grade, until the user enters 'q'. A program can be written as follows:

```
[grade_while.py]
bExit = False
while not bExit:
    command = raw_input('Enter a score ("q" to exit):')
    if command == 'q':
        bExit = True
    else:
        score = float(command)
        if 85 <= score <= 100:
            print 'A'
        elif 70 <= score < 85:
            print 'B'
        elif 50 <=score < 70:
            print 'C'
        elif 0 <= score < 50:
            print 'D'
        else:
            print 'Invalid score entered'
```

grade_while.py is an extension of *grade.py*, using a *while* loop to repeatedly perform grade calculation. The loop is controlled by a Boolean variable *bExit*, which is initialized to *False*. At each loop iteration the program asks the user to enter a score, or enter 'q' in order to quit the program. If the user enters 'q', *bExit* is set to *True*, so that the next time the *while* line is executed, the loop terminates; otherwise the original functionality of *grade.py* is performed. An execution of *grade_while.py* can give the following result:

```
Zhangs-MacBook-Pro:code yue_zhang$ python
 grade_while.py
Enter a score ("q" to exit): 100
A
Enter a score ("q" to exit): 85
A
Enter a score ("q" to exit): 90
A
Enter a score ("q" to exit): 75
B
Enter a score ("q" to exit): 55
C
Enter a score ("q" to exit): q
Zhangs-MacBook-Pro:code yue_zhang$
```

grade_while.py embeds a nested *if* structure in the *while* loop in order to combine the functionalities of repetition and grade calculation. However, the nested statements can be difficult to understand. To make the program easier to understand and maintain, the grading functionality can be taken out and put into a specific function:

```
[grade_def.py]
def grade(score):
   if 85 <= score <= 100:
      return 'A'
   elif 70 <= score < 85:
      return 'B'
   elif 50 <=score < 70:
      return 'C'
   elif 0 <= score < 50:
      return 'D'
   else:
      return None

# main
bExit = False
while not bExit:
   command = raw_input('Enter a score ("q" to exit):')
   if command == 'q':
      bExit = True
   else:
      score = float(command)
      g = grade(score)
      if g == None:
         print 'Invalid score entered'
      else:
         print g
```

grade_def.py defines a function, *grade*, which takes an input score and returns the corresponding grade. If the score is out of the range [0, 100], the special value *None* is returned. This is an explicit return statement with the *None* object as the return value, which makes use of the 'nothing' sense of the *None* object.

The rest of *grade_def.py* is a loop that repeatedly asks the user for a score, exiting if the input is the string 'q', or calculating the grade otherwise. In the latter case, the score that the user enters is passed to a *grade* function call to obtain the corresponding grade. If the grade is *None*, an error message is printed out; otherwise the grade value is printed. An execution of *grade_def.py* yields the same result as *grade_while.py*.

The use of the *grade* function not only makes the *while* loop easier to understand, but also makes the grading functionality easier to maintain. For example, if the grade boundaries are adjusted, or a new grade 'F' is introduced, a change to the grading function is sufficient, without the programmer having to worry about how the grade functionality is used.

A final note on the program above is that it consists of only three top-level statements. The first is a *def* statement, which defines a function object, and binds it with the identifier *grade*. Again, it is worth noting that a function definition is different from a function call—the *def* statement makes a new function object in memory, and associates it with a function name identifier, but never invokes the function body. The second statement in *grade_def.py* is an assignment statement that initializes *bExit*. The third statement is a compound *while* statement, with a statement block that contains calls to the *grade* function.

For another example that shows the benefit of modularity, consider again *simpson.py*. *simpson_lambda.py* uses a *lambda* function to represent the function to be integrated, so that hard-coded square values in *simpson.py* can be replaced with calls to the *lambda* function. The following program, *simpson_def.py*, takes a step further, and modularizes the functionality of integration using Simpson's rule as a function. The function *simpson* takes as input arguments the function to be integrated f, the range (a, b), and the number of sub-areas N, returning the integral $\int_a^b f(x)dx$ by Simpson's rule.

```
[simpson_def.py]
def simpson(f, a, b, n):
    d=float(b-a)/n
    #f(a)
    fa=f(a)
    #f(b)
    fb=f(b)
    # sum f(a+id)
    s1=0
    i=1
    while i<=n-1:
        s1 += f(a+i*d)
        i += 1
    # sum f(a+(i+0.5)d)
    s2=0
    i=0
    while i<=n-1:
```

```
        s2 += f(a+(i+0.5)*d)
        i += 1
    # area
    s = (d/6) * (fa+2*s1+4*s2+fb)
    return s

print 'The integral is equal to', simpson(lambda x:
    x*x, 0, 10, 10000)
```

simpson_def.py consists of two top-level statements. The first is a *def* statement that defines the function object for Simpson integration, assigning it to the identifier *simpson*, and the second is a function call to *simpson*, with the input arguments f, a, b and N being given the same values as in *simpson.py*. Note that the *lambda* function expression *lamdba* x: $x * x$ is passed as the first argument.

The advantage of *simpson_def.py* is that the functionality of Simpson integration can be reused by calls to the *simpson* function, with arbitrary functions and ranges. In fact, *simpson_def.py* can be turned into a library by commenting out the function call below:

```
# print 'The integral is equal to', simpson(lambda x:
    x*x, 0, 10, 10000)
```

simpson_test.py imports *simpson_def.py*, and calculates the integrals $\int_0^{10} x^3 dx$, $\int_0^1 sin(x)dx$ and $\int_1^{1.5} log(x)dx$, respectively.

```
[simpson_test.py]
import simpson_def
import math
print 'The integral of x^3 between 0 and 10 is',
    simpson_def.simpson(lambda x:x*x*x, 0, 10, 10000)
print 'The integral of sin(x) between 0 and 1 is',
    simpson_def.simpson(math.sin, 0, 1, 10000)
print 'The integral of log(x) between 1 and 1.5 is',
    simpson_def.simpson(math.log, 1, 1.5, 10000)
```

To calculate the integrals for $sin(x)$ and $log(x)$, *simpson_test.py* passes the function objects associated with *math.sin* and *math.log* to the function *simpson*, respectively. Take *math.sin* for example, when passed as one input argument to *simpson*, the assignment *f=math.sin* is executed before the execution of the function body, leading to the parameter f inside the function being bound to the function object *math.sin*. Hence when f is called inside the *simpson* function, the *math.sin* function object is invoked. More details on the mechanism of function calls is introduced in the next section. An execution of *simpson_test.py* yields the following result.

```
Zhangs-MacBook-Pro:code yue_zhang$ python simpson_test
    .py
The integral of x^3 between 0 and 10 is 2500.0
The integral of sin(x) between 0 and 1 is
    0.459697694132
The integral of log(x) between 1 and 1.5 is
    0.108197662162
```

6.3 Identifier Scopes

Identifier scope refers to the accessibility of identifiers in different parts of Python code; it is an important concept related to functions. For a simple example of the concept of scopes, consider the accessibility of identifiers from a function body. Statements defined in a function can contain references to identifiers defined outside the function:

```
>>> import math
>>> def f(x):
...     y = x+1
...     return math.factorial(y)
...
>>> f(1)
2
```

In the example above, the identifier *math* is defined by an *import* statement outside the function f, but the function body of f contains a direct reference to *math*. When the function call $f(1)$ is evaluated, the *math* module object is accessed by the function body successfully, yielding the return value $2! = 2$.

Global scope and local scope. Top-level statements in IDLE or a Python program being executed form the **global scope**, while statements inside a function body being executed form a **local scope**. Correspondingly, parameters (e.g. x in the example above) and identifiers defined inside functions (e.g. the variable y in the example above) are called ***local*** identifiers. Python allows global identifiers to be accessed by local statements, but not vice versa.

```
>>> x=1
>>> def f(a):
...     y=a+x
...     return y*y
...
>>> f(1)
4
>>> a
Traceback (most recent call last):
  File "<stdin>", line 1, in <module>
NameError: name 'a' is not defined
>>> y
Traceback (most recent call last):
  File "<stdin>", line 1, in <module>
NameError: name 'y' is not defined
```

In the example above, statements in the function body of f have access to the global variable x. However, the local variables a and y in the function f are not accessible by the global scope. This is intuitive, because every time the function f is executed, its local variables are assigned different values. In fact, local scopes are temporary. They are only available during the execution of a function body. Each time a function call is evaluated, a local scope is created, containing the specific arguments and variable values for the call. After the function body is executed, the local scope will become obsolete.

Note also that the global identifier x can be defined after the definition of f; as long as it is defined in the global scope *before* the first call to f, it can be accessed from the local scope when f is called.

```
>>> def f(a):
...     y=a+x
...     return y*y
...
>>> x=1
>>> f(1)
4
```

The code above runs successfully even though x is defined after the *def* statement, because the *function definition* of f does not invoke execution of the function body, and x is defined before the evaluation of the *function call* $f(1)$. On the other hand, if x is defined after the call to f, it will not be available in the global scope when the function body of f is executed.

```
>>> def f(a):
...     y=a+x
...     return y*y
...
>>> f(1) # error
Traceback (most recent call last):
  File "<stdin>", line 1, in <module>
  File "<stdin>", line 2, in f
NameError: global name 'x' is not defined
>>> x=1
```

Two local scopes. Identifiers defined in two different local scopes do not clash with each other:

```
>>> a=1
>>> def f(x):
...     y=x+a
...     z=g(y)
...     return z
...
>>> def g(x):
...     y=x*x*a
...     return y
...
>>> a=f(1)
>>> print a
```

Table 6.1 Example of nested function calls.

main	f	g
a=1		
def f(x): ...		
def g(x): ...		
a=f(1)		
	x=1	
	y=x+a	
	z=g(y)	
		x=2
		y=x*x*a
		return y
	return z	
print a		

The dynamic execution sequence of the code above is shown in Table 6.1. Five top-level statements are written in IDLE, and executed in the static order. When the fourth statement, $a = f(1)$, is executed, the right hand side of the = symbol is evaluated, and the control flow enters the function body of f, resulting a new local scope, in which the assignment of the input argument x and the three local statements are executed. When the second local statement, $z = g(y)$, is executed, the right hand side of = is evaluated, and the control flow further enters the function body of g, resulting in a new local scope, in which the assignment of the input argument x and in g the two local statements are executed. During the execution of g's function body, the local variable x is assigned the value 2, which is different from the local variable x in f—they are independent of each other, each living in its own scope.

After the function body of g is executed, the control flow goes back to f, and the assignment statement $z = g(y)$ finishes with the local identifier z being assigned the value of $g(y)$, 4. At this point, the local scope for the g call is obsolete. After the function body of f is executed, the control flow further goes back to IDLE, and the assignment statement $a = f(1)$ finishes with the global identifier a being assigned the value of $f(1)$, 4. At this point, the local scope for the f is obsolete.

The uniqueness of a function call control flow is that the dynamic execution leaves the **caller** when the function call expression is evaluated, and does not return to the caller until the **callee** has finished its own execution sequence. As a result, nested function call result in layered execution sequence such as the one in Table 6.1. This structure is also called a **stack**, which will be discussed further in Chaps. 7 and 10.

Statements in two different local scopes cannot access identifiers defined in each other. For example:

```
>>> a=1
>>> def f(x):
...     y=1
...     z=g(x+1)+y
```

```
...     return z
...
>>> def g(x):
...     return x*(y+a)
...
>>> f(1)
Traceback (most recent call last):
  File "<stdin>", line 1, in <module>
  File "<stdin>", line 3, in f
  File "<stdin>", line 2, in g
NameError: global name 'y' is not defined
```

The function call $f(1)$ above leads to an error because the reference to the identifier y in g is unresolvable. There is a y defined in f's scope, and g is called from f. However, the local scopes of f and g cannot access each other.

Nested local scope. One way of enabling g to access f's scope is making g an *internal* function of f, by putting the *def* statement of g inside f's function body:

```
>>> a=1
>>> def f(x):
...     def g(x):
...         return x*(y+1)
...     y=1
...     z=g(x+1)+y
...     return z
...
>>> f(1)
5
```

In the example, the function g is itself a local variable in f, because it is defined in the function body of f. The identifier g is in the same scope as the identifiers y and z. When g is internal to f, the scope of g is an ***inner scope*** of f. The correlation between g's scope and f's scope is similar to that between f's scope and the global scope—when Python cannot find an identifier in the current scope, it searches the ***outer scope***. Note that the up-tracing is transitive: if the identifier a can be found in neither g's scope and its *immediate outer scope*, f's scope, Python goes one level further up to find a in the global scope. As introduced earlier, the outer scope of the global scope is the *__builtins__* module scope, which Python will search if the global scope does not contain an identifier. In case the identifier is not built-in either, Python will report a name error.

When scopes are nested, an identifier in an inner scope overrides identifiers with the same name in an outer scope. For example, when the code above is executed, the value of the local identifier x in g is 2, although the value of the identifier x in f is 1. g's references to the identifier x are resolved to its local identifier.

Now suppose that the identifier a is also defined in f:

```
>>> a=1
>>> def f(x):
...     def g(x):
...         return x*(y+a)
...     y=1
...     a=0
```

```
...     z=g(x+1)+y
...     return z
...
>>> f(1)
3
```

By default, g's local reference to the identifier a will be resolved to the identifier a in the scope of f, because the immediate outer scope of g is f. To explicitly declare that g's local reference to a should be resolved to the identifier a in the global scope, the **global statement** should be used. The syntax of the statement is:

```
global <identifier>
```

The *global* statement declares that the identifier in the statement is global; it can be used only inside function definitions. Using the *global* statement, the function g above can be modified as follows:

```
>>> a=1
>>> def f(x):
...     def g(x):
...         global a
...         return x*(y+a)
...     y=1
...     a=0
...     z=g(x+1)+y
...     return z
...
>>> f(1)
5
```

The functions f and g above can be difficult to understand and expensive to maintain because of duplicated uses of the same identifier names in different **nested scopes**. In practice, the advice is to avoid name clashes wherever possible, by using reasonably long and expressive identifier names. The examples above are nevertheless useful for showing how to avoid common pitfalls and write better Python programs. For a deeper understanding of the reason behind the observations above, it is necessary to illustrate the underlying mechanisms of function definition and execution in Python.

6.4 The Underlying Mechanism of Functions

To begin with, consider the following three examples:

1. Example (1): local variable.

```
>>> a=0
>>> def f(x):
...     a=1
...     print 'a in f:', a
...
>>> f(1)
```

```
a in f: 1
>>> print 'a in global:', a
a in global: 0
```

2. Example (2): global variable.

```
>>> a=0
>>> def f(x):
...     global a
...     a=1
...     print 'a in f:', a
...
>>> f(1)
a in f: 1
>>> print 'a in global:', a
a in global: 1
```

3. Example (3): incorrect variable.

```
>>> a=0
>>> def f(x):
...     print 'a in f:', a
...     a=1
...
>>> f(1)
a in f:
Traceback (most recent call last):
  File "<stdin>", line 1, in <module>
  File "<stdin>", line 2, in f
UnboundLocalError: local variable 'a' referenced
    before assignment
>>> print 'a in global:', a
a in global: 0
```

The examples (1) and (2) above are straightforward to understand; the example (3) gives an error because the variable a in f is local (no *global* declaration), but referred to before being assigned a value. Python reports an *unbounded error* rather than a *name error* in this case, and the reason can be understood with more details of the underlaying mechanism of function call execution.

In memory, a function object maintains three main sources of information by keeping

1. a list of local and global identifiers (including parameters),
2. the code block, which is a list of statements from the function body, and
3. a pointer to the binding table that corresponds to the immediate outer scope of the function.

When a *def* statement is executed, Python creates a function object, filling it with all the information above, and associates it with the given identifier in the current binding table. Figure 6.2 gives the memory structures of the examples (1)–(3) above, after the function definition statements are executed, but before the function calls are executed. For both the examples (1) and (3) above, the identifier a is local, but for the example (2), the identifier a is global. In all the three examples, the current scope

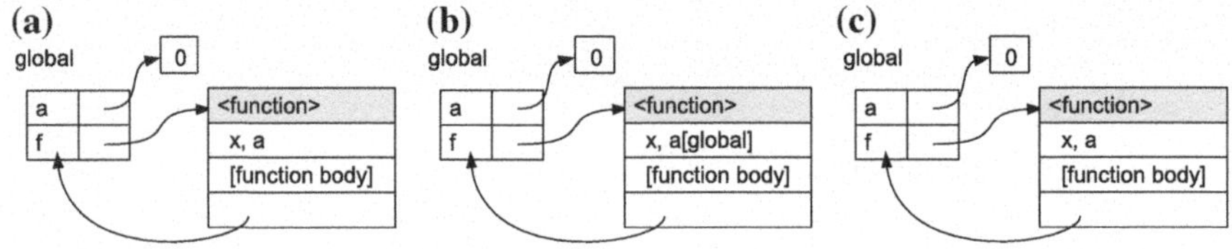

Fig. 6.2 Memory structure of function objects

is the global scope when the *def* statement is executed, and therefore the variable *f* is created in the global binding table.

The underlying mechanism of a function call consists of three main steps:

1. Creating a binding table for the local scope. The binding table contains all the local identifiers (including input parameters), and has a pointer to the immediate outer scope. It is constructed by initializing an empty binding table, filling it with the list of local identifiers kept by the function object, and creating a link to the outer binding table according to the outer-scope pointer kept by the function object. Figure 6.3 gives the memory structures of the examples (1)–(3) above, when this step of the fucntion call is executed.
2. Binding parameters with the input argument objects specified in the function call. This step corresponds to the argument assignment step in function execution.
3. Executing the function body, with the newly constructed binding table being the current local scope. The execution of each statement in the function body follows Python's mechanism of the respective statement. In partifular, when an identifier is evaluated, the current binding table is searched; in case there is not an entry for the identifier, outer binding tables are searched in order by following the outer-scope pointers. Figure 6.3 illustrates the binding table hierarchy by showing the *__builtins__* scope explicitly.

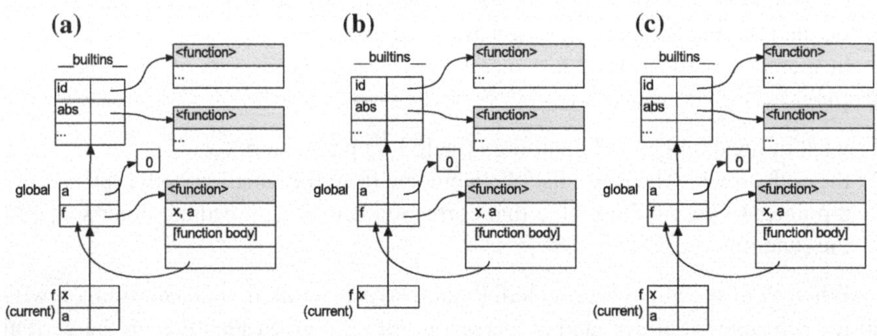

Fig. 6.3 Mechanism of function execution. **a** Example (1), **b** Example (2), **c** Example (3)

The error in the example (3) above can be explained by Fig. 6.3c. Before the function body is executed, the argument x has been associated with the object 1, but the local identifier a has not been initialized. The first statement to execute is the *print* statement, which contains a reference to a in the string expression, and thus results in the unbounded erorr above.

For another example of memory structures, consider the execution of *simpson_test.py* in the previous section. Figure 6.4 shows the memory structures at different stages of execution. In particular, Fig. 6.4a shows the memory structure after the two *import* statements are executed. The mechanism of module import has been introduced in Chap. 3; when each *import* statement is executed, the corresponding module is executed, and their binding tables are associated with respective identifiers in the binding table of the importer.

Figure 6.4b shows the memory structure after the third statement has been executed. The statement contains a function call expression, which is evaluated by invoking the function *simpson*. In this process, the three basic steps of a function call are carried out, with a new binding table being constructed, all arguments being assigned to their corresponding parameters, and the function body being executed using the new binding table as the current binding table. After all the statements in the function body are executed, the current binding table is switched back to the global binding table for the completion of the *print* statement, and the local binding table of the function that has just been executed is left as shown in the figure. Because it is no longer referred to directly or indirectly by the current binding table, the local binding table will be removed by the garbage collector automatically.

Figure 6.4c shows the memory structure when the fourth statement is being executed, during the evaluation of the function call to *simpson*, just after the arguments are assigned, but before the function body is executed. In this incremental process, a new binding table is constructed and filled with the local variables of *simpson*. This binding table is different from the one created by the third statement, because each function call leads to a new local binding table, independent of other calls to the same function. The next step will be the execution of the function body, with the new binding table being taken as the current binding table.

For a final example of memory structures, consider the factory function *twice*, which takes an input function $f \in \mathcal{R} \rightarrow \mathcal{R}$ and returns a function that is equivalent to $f \circ f$.

```
>>> def twice(f):
...     def g(x):
...         return f(f(x))
...     return g
...
>>> import math
>>> f1=twice(math.sqrt)
>>> f1(625)
5.0
>>> f2=twice(lambda x : x+1)
>>> f2(1)
3
```

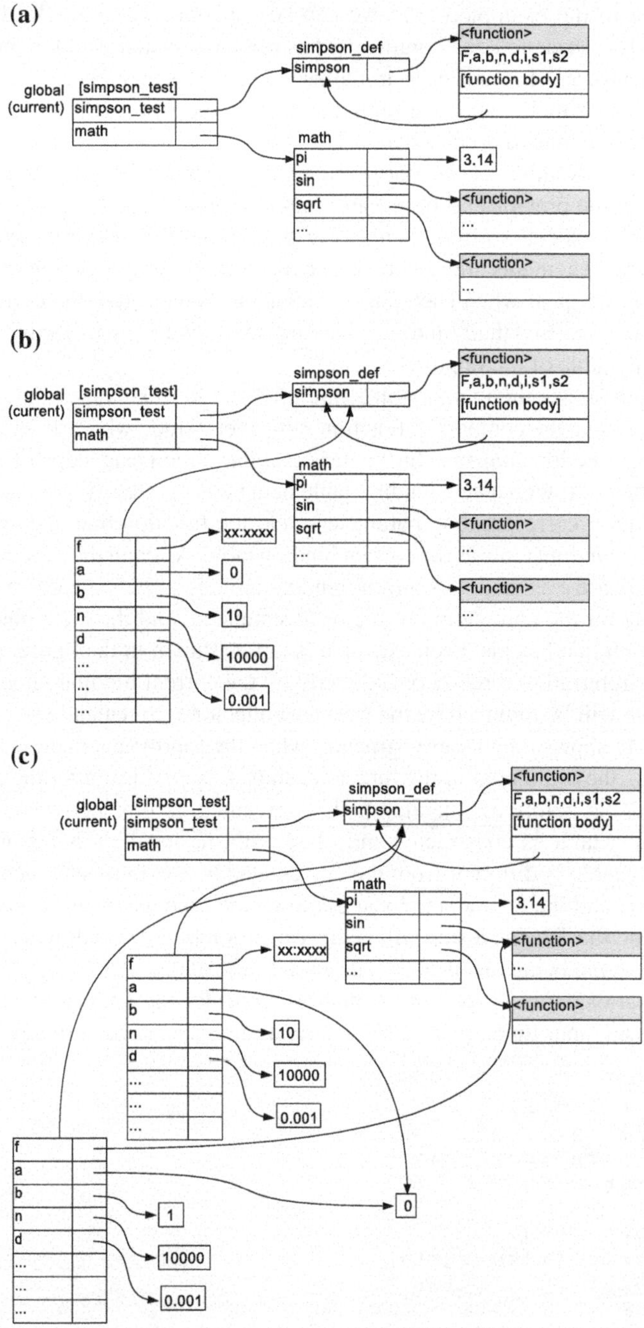

Fig. 6.4 Memory structures during the execution of *simpson_test.py*. **a** After the first two statements are executed, **b** After the first three statements are executed, **c** When the fourth statement is being executed, just after argument assignment in the simpson call

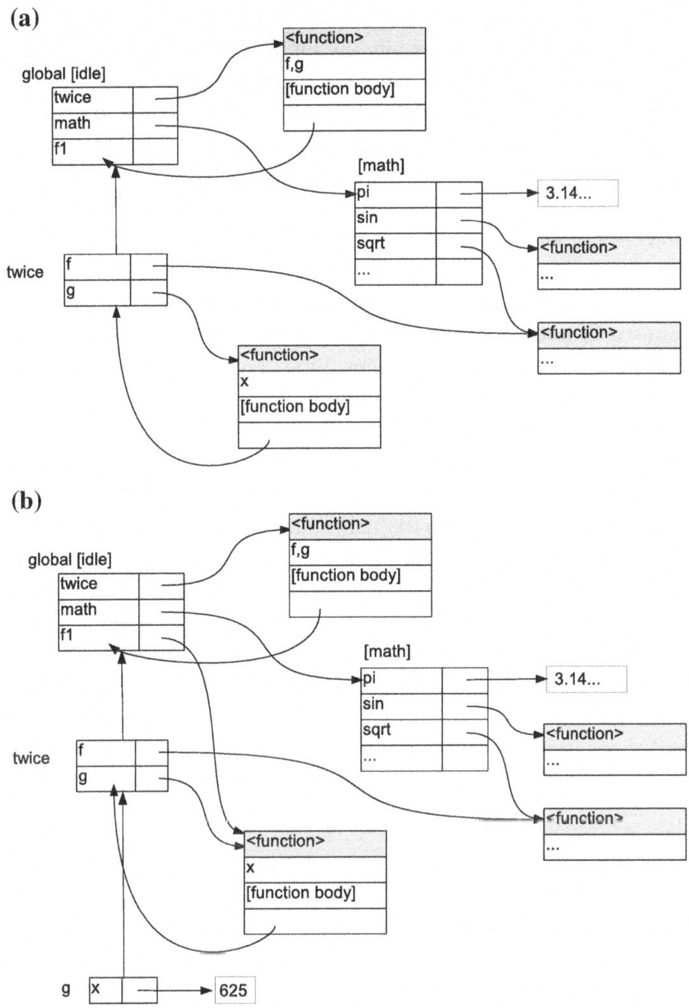

Fig. 6.5 Memory structures during the execution of *twice*. **a** After *f1*=twice*(math.sqrt)* is executed, **b** During the execution of *f1(625)*

In the example above, the function *twice* has an internal function, g, which takes an input argument x and returns $f(f(x))$, where f is the input function to the function *twice*. The function *f1* applies *math.sqrt* twice, and the function *f2* applies *lambda* x:$x + 1$ twice.

The memory structure after the statement *f1=twice(math.sqrt)* is executed as shown in Fig. 6.5a. The function call *twice(math.sqrt)* is executed in three steps. First, a new binding table is constructed and filled with the local variables f and g. Second, the parameter f is bound to the input argument value *math.sqrt*. Third, the function body is executed. The first statement in the function body is a function definition statement, which creates a new function object, with a single local variable x, and an outer-scope pointer to the binding table of *twice*. The new function object is assigned to the local identifier g in the binding table of *twice*.

The second statement in the function body of *twice* is a *return* statement, with the return value being g. Execution of this statement results in the function call expression *twice(math.sqrt)* being evaluated to the function object of g, which is assigend to the identifier $f1$ in the global binding table. After the execution of this statement, the function object of g will not be erased by the garbage collector, because it is referred to by the global identifier $f1$. Neither will the local binding table of *twice* be erased, because it is referred to by the outer-scope pointer of the function object of g.

The memory structure when the call $f1(625)$ is executed is shown in Fig. 6.5b. The execution of $f1$ consists of three steps. First, a new binding table is constructed, with one entry x. Second, the local identifier x is bound to the input integer object 625. Third, the function body is executed. There is only one statement in the function body of $f1$, *return f(f(x))*, which refers to the identifiers x and f. In the execution of this statement, x can be found in the local binding table, but f is missing. Python follows the outer-scope pointer up to the binding table of *twice*, where the identifier f is found, referring to the function object *math.sqrt*.

When the statement *f2=twice(lambda x:x+1)* is executed, the function *twice* is called again, with a new local binding table being constructed, and a new local funciton object g being created, whose outer-scope pointer points to the local binding table of this call to *twice*. The return value g is then assigned to the identifier *f2* in the global binding table. Because the outer scope of *f2* is the binding table of the second execution of *twice*, with the input argument f being *lambda x: x+1*, the call to *f2* results in two nested calls to the *lambda* function, rather than the function *math.sqrt*.

Exercises

1. Which of the following programs are correct? Why?

   (a) returns the first power of 3 in a tuple.

   ```
   def f(t):
       for x in t:
           if x%3 == 0:
               break
   ```

   (b) returns all powers of 3 in a tuple.

   ```
   def f(t):
       for x in t:
           if x%3 == 0:
               return x
   ```

   (c) a function that takes a number argument and returns '+' when the argument is positive, and '-' if the argument is negative.

   ```
   def posneg(x):
       if x>0:
           return '+'
       else:
           return '-'
   ```

(d) a function that takes a number argument and returns '+' if the argument is positive, '-' if the argument is negative, and '0' otherwise.

```
def posneg(x):
    if x>0:
        return '+'
    if x<0:
        return '-'
    return '0'
```

(e) a function that takes a number argument x and returns 'A' if $x > 10$, 'B' if $0 \leq x \leq 10$, and 'C' if $x < 0$.

```
def f(x):
if x>10:
    return 'A'
if x>=0:
    return 'B'
return 'C'
```

(f) a function that takes a number argument x and returns 'A' if $x > 10$, 'B' if $0 \leq x \leq 10$, and 'C' if $x < 0$.

```
def f(x):
if x<0:
    return 'C'
if x>=0:
    return 'B'
if x>10:
    return 'A'
```

2. What is the output of the following program?

```
import math
def f(x):
        return x*x*x
t=(math.sin, math.sqrt, lambda x: x+1, f)
for g in t:
        print g(1.0)
```

3. Write a function *f*, which takes a function *g* and up to two additional arguments, returning the return value of a call to *g* with the other arguments being passed as its arguments. For example,

```
>>> f(math.sqrt, 25)
5.0
>>> f(math.pow, 5.0, 2)
25.0
```

4. write a function *mol*, which can take either one, or two, or three arguments, returning their product in *float* type. For example, *mol*(1.0) will return 1.0, *mol*(1.0, 1.5) will return 1.5, and *mol*(2.0, −1.0, 3.0) will return 6.0.

5. Write a function for the following tasks.
   (a) turn the Monte-Carlo method for numerical integration, discussed in the previous chapter, into a function;
   (b) Write a function *argmax*(l), which takes a single tuple argument, and return a tuple that consists of the maximum item in the type and its index. For example, *argmax*((1, 3, 2, −1, 4, 6, −7, −8)) will return (6, 5);
   (c) write a function *prime*(x), which takes an integer argument x and returns a Boolean value indicating whether x is a prime number;
   (d) write a function *factors*(x), which takes an integer argument x and returns a tuple that contains all the factors of x;
   (e) write a function *gcd*(x,y) that takes two integer arguments x and y, and returns their greatest common divisor.
6. A leap year is a year that can be divided by 4, but not 100. Write a function *leap-year* that does not take any input argument, and asks the user to input a start year and an end year, outputting a tuple that contains all the leap year between the start year and the end year, inclusive. For example, one execution of print *leap-year*() can result in the following output.

   Enter start year: 1998

   Enter end year: 2009

   (2004, 2008)

Chapter 7
Lists and Mutability

All the types that have been introduced up to this chapter, including the number types in Chap. 2, the string type in Chap. 3, the Boolean type in Chap. 4, and *tuple* type in Chap. 5 and the *NoneType* in Chap. 6, are **immutable types**. Once created, *objects* that belong to immutable types cannot change their *values*.[1]

This chapter introduces the first **mutable type: lists**. Objects that belong to mutable types can be changed after they are created. As the result, the value of a variable that is bound to a mutable object can be changed without using an assignment statement (i.e. changing the binding). On the other hand, once the value of an object changes, all the identifiers that are bound to the object are affected, changing their values accordingly. One can take advantage of this fact and modify a set of variables simultaneously. However, programming mistakes can also be introduced by overlooking this fact.

7.1 Lists—A Mutable Sequential Type

The **list** type is the third sequential type introduced in this book. It is a container type, and can be regarded as a mutable version of tuples. A **list literal** consists of a comma-separated list of objects and identifiers, enclosed in a pair of square brackets.

---

[1]The values of *variables* that are bound to immutable objects can change, because the assignment statement can change the binding between identifiers and objects. For example, when the statement $x = x + 1$ or $x += 1$ is executed, the identifier x is unbound from its previous value o_1 and bound to a new object o_2, the value of which is $o_1 + 1$. In this process, neither the object o_1 nor the object o_2 changes, but the value of the variable x changes. In memory, the assignment statement changes only the binding table, but not the number objects.

Y. Zhang, *An Introduction to Python and Computer Programming*,
Lecture Notes in Electrical Engineering 353, DOI 10.1007/978-981-287-609-6_7

```
>>> type([1,2,3])
<type 'list'>
>>> l=[1]                  # one item
>>> type(l)
<type 'list'>
>>> l
[1]
>>> x='a'
>>> l=[x, True, None]      # heterogeneous list
>>> type(l)
<type 'list'>
>>> l
['a', True, None]
```

List literals are different from tuple literals only in the enclosing brackets (i.e. [] instead of ()). As shown by the example above, unlike tuple literals, no comma is needed in a list literal that contains only one item, because unlike round brackets, the use of square brackets is unambiguous.

List Operators. The *list* type supports all the operations of strings and tuples, including +, *, *in*, *not in*, *getitem* and *getslice*. In addition, the *len* function can also be applied to lists.

```
>>> l=[1, 'a', False]
>>> l + [1.0, 1+2j]
[1, 'a', False, 1.0, (1+2j)]
>>> l * 2
[1, 'a', False, 1, 'a', False]
>>> 'a' in l
True
>>> 'b' in l
False
>>> 1 not in l
False
>>> 2 not in l
True
>>> l[0]    #getitem
1
>>> l[0:2] #getslice
[1, 'a']
>>> l[-1]
False
>>> l[-2:]
['a', False]
>>> len(l)
3
>>> len([]) # empty list
0
```

Conversion Between Lists and Other Types. A list object can be converted to a string, a Boolean object, or a tuple object. However, it cannot be converted to numbers. Similar to *tuples*, the string conversion of a list object is its literal form.

```
>>> x='a'
>>> y=1
>>> z=True
```

```
>>> l=[x, y, z]
>>> str(l)
"['a', 1, True]"
>>> int(l) # cannot be converted numbers
Traceback (most recent call last):
  File "<stdin>", line 1, in <module>
TypeError: int() argument must be a string or a number
    , not 'list'
```

Similar to the cases of tuples and strings, the Boolean conversion of a *list* is *False* only when the *list* is empty, and *True* otherwise.

```
>>> bool([])
False
>>> bool([1,2,3])
True
>>> l=[None]
>>> bool(l)
True
```

The tuple conversion of a *list* object is a *tuple* object that contains exactly the same elements in the same order.

```
>>> l=[1+2j, 'abc', None]
>>> t=tuple(l)
>>> t
((1+2j), 'abc', None)
```

On the opposite direction, string and *tuple* objects can be converted into *list* objects. However, Boolean or number objects cannot be converted into *list* objects. Similar to the *tuple* conversion, the *list* conversion of a string is a *list* that consists of all the characters in the string, each as a *list* item, in their original order. Conversion from *tuples* to *lists* is the reverse operation to the conversion from *lists* to *tuples*.

```
>>> l=[1+2j, 'abc', None]
>>> t=tuple(l)
>>> t
((1+2j), 'abc', None)
>>> list("hello!")
['h', 'e', 'l', 'l', 'o', '!']
>>> list((1, 7.5, 1+2j, 'abc', True, None))
[1, 7.5, (1+2j), 'abc', True, None]
>>> list(True) # cannot be converted from Boolean
    objects
Traceback (most recent call last):
  File "<stdin>", line 1, in <module>
TypeError: 'bool' object is not iterable
>>> list(1.0) # neither numbers
Traceback (most recent call last):
  File "<stdin>", line 1, in <module>
TypeError: 'float' object is not iterable
```

7.1.1 List Mutation

There are two ways to make changes to a *list* object. The first is *setitem* and *setslice*, which directly replace an item or a slice in a list object, respectively, and *delitem* and *delslice*, which delete an item or a slice from a list object, respectively.

The syntax of the *setitem* operation takes the form of the assignment statement—the = symbol is used for assigning the value on its right hand side to a 'variable' on its left hand side. The only difference is that for the *setslice* operation, the 'variable' on the left hand side is an item from a *list* instead of an identifier.

```
>>> l=[1,2,3]
>>> l[0]=0
>>> l
[0, 2, 3]
>>> l[-1]=10
>>> l
[0, 2, 10]
```

As shown by the examples above, the syntax for specifying a *list* item in the *setitem* operation is the same as that in the *getitem* operation: an integer in a pair of square brackets is used to indicate the index of the item. The meaning of the = symbol for the *setitem* operation is consistent with the assignment statement (i.e. assigning a value to an identifier). In the *setitem* operation, the *list* item plays the role of an identifier, the value of which changes according to the assigment.

In memory, a list object can be treated as a sequence of binding slots, as shown on the right hand side of Fig. 7.1. The functionality of = in both variable assignment and *setitem* is to change bindings. The only difference is that, for the former, the change is in a binding table, but for the latter, the change is in a *list* object.

Suppose that two identifiers are bound to the same *list* object. Once the *list* object is modified, the value of both identifiers changes.

```
>>> x=['a', 'b', 'c']
>>> y=x
>>> y[0]='A'
>>> y
['A', 'b', 'c']
>>> x
['A', 'b', 'c']
```

Figure 7.1 illustrates the changes in memory when the *setitem* operation is performed. Before $y[0] =$ '*A*' is executed, the identifiers x and y in the global binding table point to the same list object. $y[0] =$ '*A*' modifies the first item of the *list* object without affecting the global binding table. As a result, the value of both x and y is modified.

The *setslice* operation has the same syntax as the *setitem* operation, except that the left hand side of the = symbol specifies a slice from a *list* object rather than an item, and the right hand side of the = symbol must be convertable to a *list* object.

```
>>> l = [1, 2, 3]
>>> l[0:2] = [-1, -2] # replaces a slice
>>> l
```

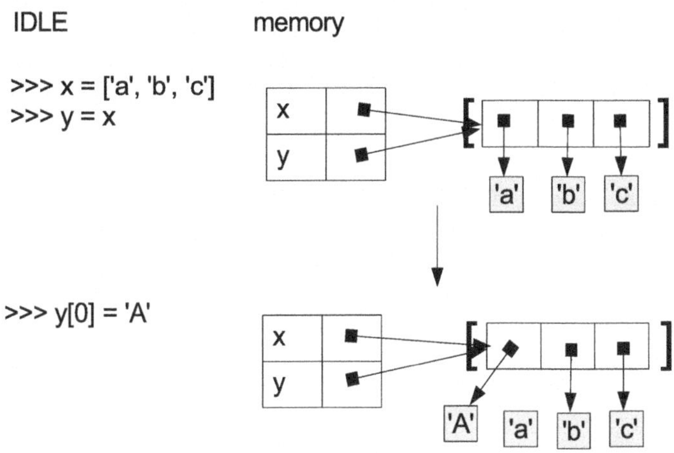

Fig. 7.1 Memory structure for shared list modification

```
[-1, -2, 3]
>>> l[1:3] = [2, 3, 4] # replaces a slice with a
    longer slice
>>> l
[-1, 2, 3, 4]
>>> l[-2:] = [-3] # replaces a slice with a shorter
    slice
>>> l
[-1, 2, -3]
>>> l[1:2] = [] #deletes a slice
>>> l
[-1, -3]
>>> l[1:1] = [-2] # inserts a slice
>>> l
[-1, -2, -3]
```

In the example above, the slices specified by the left hand side of the = symbols are replaced with the *lists* specified by the right hand side. Two cases are worth noting. The first is $l[1:2] = []$, where the right hand side is an empty *list* — the effect of replacing a non-empty slice with an empty list is to delete the corresponding slice. The second is $l[1:1] = [-2]$, where the left hand side is an empty slice — the effect of replacing an empty slice with a non-empty list is to insert a *list* into the location specified by the empty slice. As introduced in Chap. 3, the starting index of a slice indicates the location before the corresponding item.

The *setslice* operation does not create any binding relationship between the *list* on the left hand side of = and the *list* on the right hand side. As a result, two independent *lists* remain independent after a *setslice* operation between them:

```
>>> l = [1, 2, 3]
>>> s = ['a', 'b']
>>> l[2:] = s
>>> l
```

```
[1, 2, 'a', 'b']
>>> l[3] = 'c'
>>> l
[1, 2, 'a', 'c']
>>> s
['a', 'b']
```

In the example above, the slice $l[2:]$ was set to the value of s. After this operation, l is further changed. The change $l[3] =$ 'c' modifies the new slice on l, but does not change the value of s. The reason behind can be explained by Fig. 7.2, which shows the mechanism of *setslice*. The new slice in e is a copy of s, rather than s itself. In fact, the right hand side for *setslice* can also be immutable sequences, such as *tuples* and strings.

```
>>> l = [1, 2, 3]
>>> t = (1.0, 2.0, 3.0)
>>> s = 'abc'
>>> l[2:] = t
>>> l
[1, 2, 1.0, 2.0, 3.0]
>>> l[4:] = s
>>> l
[1, 2, 1.0, 2.0, 'a', 'b', 'c']
```

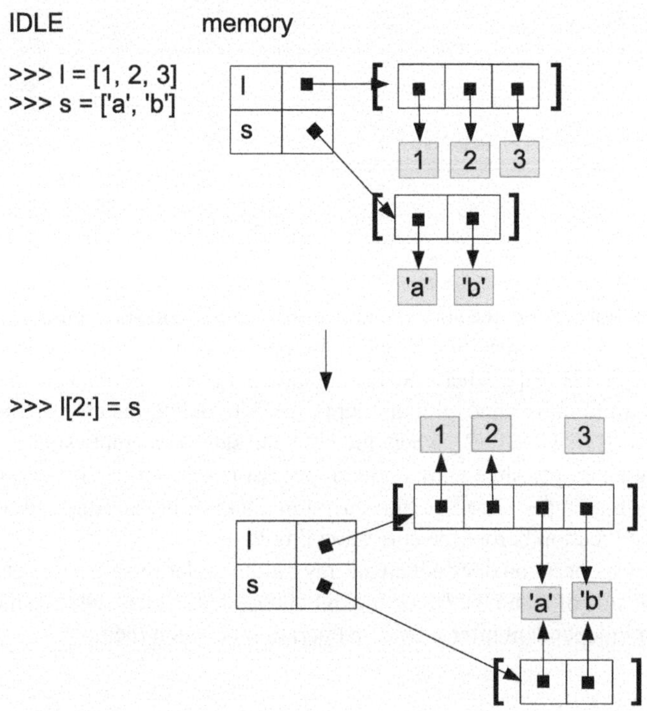

Fig. 7.2 Memory structure for the *setslice* operation

The reverse operation of the assignment statement is the *del* statement, as introduced in Chap. 2. *del* can also be applied to *list* items and slices, indicating the *delitem* and *delsice* operations, respectively.

```
>>> l = [1, 2, 3, 4, 5, 6, 7, 8, 9]
>>> del l[0] # removes the first item
>>> l
[2, 3, 4, 5, 6, 7, 8, 9]
>>> del l[2:5] # removes a slice
>>> l
[2, 3, 7, 8, 9]
>>> del l[-1]
>>> l
[2, 3, 7, 8]
```

One final note about the *setitem*, *setslice*, *delitem* and *delslice* operations is that they are applicable only to *lists*, which are mutable, but not to strings or *tuples*, which are immutable.

```
>>> s='abc'
>>> t=(1,2,3)
>>> s[0]='d' # no string setitem
Traceback (most recent call last):
  File "<stdin>", line 1, in <module>
TypeError: 'str' object does not support item
    assignment
>>> t[1:3]=(0,) # no tuple setslice
Traceback (most recent call last):
  File "<stdin>", line 1, in <module>
TypeError: 'tuple' object does not support item
    assignment
>>> del s[1:] # no string delslice
Traceback (most recent call last):
  File "<stdin>", line 1, in <module>
TypeError: 'str' object does not support item deletion
>>> del t[-1] # no tuple delitem
Traceback (most recent call last):
  File "<stdin>", line 1, in <module>
TypeError: 'tuple' object doesn't support item
    deletion
```

Methods and attributes. The second way of modifying a *list* object is to call its *methods*. A **method** is a special function, which is bound to a specific object, performing operations on the object. The syntax of a *method call* is

```
<object>.<method name>(<arguments>)
```

A method call starts with an identifier, which represents the object, followed by a dot (.), and then the name of the method that is bound to the object, and finally a comma-separated list of arguments, enclosed in a pair of round brackets. The syntax of a method call can be split into two parts, the first being the object and the dot, and the second being the method name and the arguments. The second part is identical to a function call, while the first half indicates that the method belongs to a specific object.

Complex numbers have a method, *conjugate*, which takes no input arguments, and returns the conjugate of the complex object. For example:

```
>>> a=1+2j
>>> b=3+4j
>>> a.conjugate()
(1-2j)
>>> b.conjugate()
(3-4j)
```

It can be seen from the example above that methods are object-specific: the *conjugate* methods bound to the complex numbers *a* and *b* give different return values. Here the role of *a* and *b* is similar to that of an argument. For example, the built-in function *abs* takes one input argument, and returns its absolute value. It gives different return values when applied to different input arguments. If the *conjugate* method were written as a global function, taking one complex argument and returning its conjugate, *a.conjugate()* would have been written as *conjugate(a)*, with the same effect. This is similar to a call to *abs(x)*, where *x* is a number. However, since *conjugate* applies to complex numbers only, it would be more elegant to define it as a method that applies to complex objects. The definition of methods, and for underlying connection between method-object correlation and argument-function correction, are discussed in Chap. 10.

Intuitively, methods are type-dependent. The method *conjugate* applies to complex numbers, but not to *lists*, strings or other types.

```
>>> l=[1,2,3]
>>> l.conjugate() # lists do not have the conjugate
   method
Traceback (most recent call last):
  File "<stdin>", line 1, in <module>
AttributeError: 'list' object has no attribute '
   conjugate'
>>> s='abc'
>>> s.conjugate() # strings do not have the conjugate
   method
Traceback (most recent call last):
  File "<stdin>", line 1, in <module>
AttributeError: 'str' object has no attribute '
   conjugate'
```

In addition to *methods*, which are object-specific *functions*, objects can also have **attributes**, which object-specific, but can be any type, including integers and strings. The general syntax of accessing an attribute of an object is

```
<object>.<attribute>
```

Complex numbers have two attributes, *real and imag*, which specify the real and imaginary parts of a complex number, respectively.

```
>>> a = 1 + 2j
>>> b = 3j
>>> a.real
1.0
>>> a.imag
```

```
2.0
>>> b.real
0.0
>>> b.imag
3.0
```

In the example above, the *real* attributes of the complex numbers *a* and *b* are different, and specific to the corresponding objects. Similar to methods, attitudes are type-specific, and the complex attributes *real and imag* do not apply to *lists*, *tuples* or strings. Attributes and methods are two important concepts that are related to the thinking of **object oriented programming**, which is introduced in Chap. 10.

List modification methods. There are two commonly-used methods that add an item to a *list*: *append*, which takes a single argument and adds it to the back of a *list*, and *insert*, which takes two arguments specifying an index and a new item, respectively, inserting the item into the *list*, so that the index of the new item after the insertion is the specified index. Both methods are procedures, with a *None* return value.

```
>>> l = [1, 2, 3]
>>> l.append(4)
>>> l
[1, 2, 3, 4]
>>> l.insert(0, 0)
>>> l
[0, 1, 2, 3, 4]
>>> l1 = l
>>> l.insert(3, 'a')
>>> l
[0, 1, 2, 'a', 3, 4]
>>> l1
[0, 1, 2, 'a', 3, 4]
>>> l.insert(-2, 'b')
>>> l
[0, 1, 2, 'a', 'b', 3, 4]
>>> l1
[0, 1, 2, 'a', 'b', 3, 4]
```

In the example above, the identifies *l* and *l*1 are bound to the same object. The values of *l*1 changes after the *insert* and *append* methods are applied to *l*, showing that the methods modify the *list* object itself. The effect of *append(x)* is the same as that of the *setslice* operation *l[len(l):len(l)]=x*, while the effect of *l.insert(i, x)* is the same as the *setslice* operation *l[i:i]=[x]*.

A method that extends a *list* by concatenating it with another *list* is *extend*, which takes a *list* argument, and concatenates it with the *list* object.

```
>>> l = [1, 2, 3]
>>> x = [4, 5, 6]
>>> l.extend(x)
>>> l
[1, 2, 3, 4, 5, 6]
>>> l[-1] = 0
>>> l
[1, 2, 3, 4, 5, 0]
```

```
>>> x
[4, 5, 6]
```

As the example above demonstrates, the *extend* method makes a copy of the argument x, and updates the object l with the copy of x. A modification to the new slice in l does not affect the value of x. The effect of *l.extend(x)* is the same as that of the *setslice* operation $l[len(l):len(l)] = x$.

A method that deletes an item from a *list* is *remove*, which takes one argument, and deletes the first occurrence of the argument from the *list*. If the input argument is not in the *list*, a *value error* is reported.

```
>>> l = [1, 2, 3, 4, 5, 1]
>>> l.remove(1)
>>> l
[2, 3, 4, 5, 1]
>>> l.remove(5)
>>> l
[2, 3, 4, 1]
>>> l.remove(6)
Traceback (most recent call last):
  File "<stdin>", line 1, in <module>
ValueError: list.remove(x): x not in list
```

Another method that modifies a *list* is *reverse*, which reverses the order of the *list*. The *reverse* method takes no arguments, and returns *None*.

```
>>> l = [1, 2, 3]
>>> l.reverse()
>>> l
[3, 2, 1]
```

Another method that changes the order of a *list* is *sort*, which puts items in the *list* in ascending order. Similar to *reverse*, the *sort* method takes no arguments and returns *None*.

```
>>> l = [1, 5, 3, 4, 2, 7, 6, 8, 10, 9]
>>> l.sort()
>>> l
[1, 2, 3, 4, 5, 6, 7, 8, 9, 10]
```

7.2 Working with Lists

The sequential type is the most commonly used container type in programming, and mutability offers list versatility in their use. As the result, *lists* are one of the most powerful tools in Python programming. This section discusses typical uses of *lists*, and common mistakes to avoid when working with *lists*.

7.2.1 Copying Lists

There are situations where a copy of a *list* object is needed. For example, when a modified version of a *list* is requested, but the original *list* must be kept for further uses, a copy of the original *list* can be made for the modified version. There are two ways to obtain a copy of a *list* object, namely the *list* slicing operation and the *copy* module.

The *getslice* operation for a *list* object makes a copy of a slice from the *list* object, and returns the copy rather than a reference to the original slice. For example,

```
>>> l = [1, 3, 5, 7, 9]
>>> x = l[2:4]
>>> x
[5, 7]
>>> x[0] = 0
>>> x
[0, 7]
>>> l
[1, 3, 5, 7, 9]
```

In the example above, the value of l does not change when the value of x changes, because x and l do not have direct binding relationships.

The memory structures of x and l is shown in Fig. 7.3. Note that copying a *list* or sub *list* duplicates only the *list* structure, which is a sequence of binding links, but does not duplicate the objects that are referred to by these links.

A simple way of copying a *list* is to slice the whole *list*. As introduced in Chap. 3, when both the start and the end indices are omitted in a *getslice* operation, the whole sequence is taken as the resulting slice. This provides a more succinct way of copying a *list*:

```
>>> l = [1, 3, 5, 7, 9]
>>> l1 = l[0:len(l)] # the whole list sliced and copied
>>> l2 = l[:] # a more succinct way of slicing a list
>>> l[0] = 0
>>> l1[1] = 0
>>> l2[2] = 0
>>> l
[0, 3, 5, 7, 9]
```

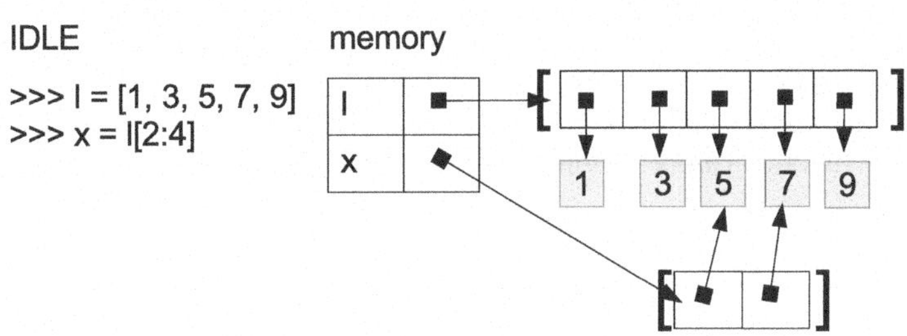

Fig. 7.3 Illustration of *getslice* for *list* objects

```
>>> l1
[1, 0, 5, 7, 9]
>>> l2
[1, 3, 0, 7, 9]
```

As shown by the example above, modifications to *l*, *l*1 and *l*2 do not affect each other, because they are different copies of the same *list* object.

A second way of making a copy of a *list* object is using the *copy* module. The module contains a function, *copy*, which takes a single argument, which is typically a mutable container object, and returns a copy of the argument.

```
>>> import copy
>>> l = [1, 3, 5, 7, 9]
>>> l1 = copy.copy(l)
>>> l[0] = 0
>>> l
[0, 3, 5, 7, 9]
>>> l1
[1, 3, 5, 7, 9]
```

For an example where *list* copying is used, consider the problem of deciding whether an integer is a palindrome number. A *palindrome number* is a number that remains the same when the order of the digits in it is reversed. For example, 12321 and 1111 are palindrome numbers, but 123 is not a palindrome number. The function *palindrome* below takes an integer argument and returns a Boolean value indicating whether the argument is a palindrome number.

```
>>> def palindrome(n):
...     l = list(str(n))
...     r = l[:]
...     r.reverse()
...     return l==r
...
>>> palindrome(12321)
True
>>> palindrome(1111)
True
>>> palindrome(123)
False
```

The *palindrome* function above works by converting the integer argument into a string, so that each digit is converted into a character. Then it converts the string into a *list* of all the digits in the integer. The digits are put into the reverse order by making a copy of the *list* and reversing it. A comparison between the reversed version and the original list determines whether the integer is a palindrome number.

Given that the slicing operation $l[::-1]$ makes a reverse copy of the *list*, the *palindrome* function can be simplified as:

```
>>> def palindrome(n):
...     l = list(str(n))
...     r = l[::-1]
...     return l==r
```

Further given that the slicing operation applies to strings also, the function can be further simplified, without using a *list*.

```
>>> def palindrome(n):
...     s = str(n)
...     r = s[::-1]
...     return s==r
```

7.2.2 Lists as Items in Tuples and Lists

A *list* can be an item in a *list* or *tuple*, in which case it is called a **sub list**.

```
>>> l=[[1,2,3],[4,5,6],[7,8,9]]
>>> l[0]                # sub list
[1, 2, 3]
>>> l[0][0]             # sub list item
1
>>> l[1][-1]            # sub list item
6
```

A *list* of *lists* is called a **nested list**. When making a copy of a nested *list*, the sub *lists* are shared by the two copies.

```
>>> l1=[[1,2,3],[4,5,6],[7,8,9],10]
>>> l2=l1[:]              # copy first level
>>> l1[3]=11              # change first level
>>> l1
[[1, 2, 3], [4, 5, 6], [7, 8, 9], 11]
>>> l2
[[1, 2, 3], [4, 5, 6], [7, 8, 9], 10]
>>> l2[2]=[10,11,12] # change first level
>>> l2
[[1, 2, 3], [4, 5, 6], [10, 11, 12], 10]
>>> l1
[[1, 2, 3], [4, 5, 6], [7, 8, 9], 11]
>>> l2[0][0]=13
>>> l2
[[13, 2, 3], [4, 5, 6], [10, 11, 12], 10]
>>> l1
[[13, 2, 3], [4, 5, 6], [7, 8, 9], 11]
```

As can be seen from the example above, a copy of a nested *list* can lead to unexpected common changes in both copies. To avoid this, the function *copy.deepcopy* can be used to copy nested lists recursively. The syntax is the same as *copy.copy*.

```
>>> l1=[[1,2,3],[4,5,6],[7,8,9],10]
>>> import copy
>>> l2=copy.deepcopy(l1)
>>> l2[0][0]=13
>>> l2
[[13, 2, 3], [4, 5, 6], [7, 8, 9], 10]
>>> l1
[[1, 2, 3], [4, 5, 6], [7, 8, 9], 10]
```

As shown in the example above, *l2* and *l1* are fully independent after *copy.deepcopy*. When copying a nested *list*, deep copying is preferred.

The underlying mechanisms of copying and deep copying. The following example is not practically useful, but can serve to illustrate the memory structures of *lists* and copying.

```
>>> l=[1,2,3]
>>> t1 = (l, 'a', True)
>>> t1
([1, 2, 3], 'a', True)
>>> l1 = [l, l, 1.0, None]
>>> l1
[[1, 2, 3], [1, 2, 3], 1.0, None]
```

In the example above, *t*1 contains three items, including a *list*, a string and a Boolean object. *l*1 contains four items, including two *lists*, a floating point number and a *None*.

The memory structures of *l*, *t*1 and *l*1 are shown in Fig. 7.4. The list object [1, 2, 3] is shared by four different references, including a global identifier *l*, the first item in the tuple *t*1, and the first and second items in the *list l*1. As a result, when it is modified, *l*, *t*1 and *l*1 are affected simultaneously.

```
(continued from above)
>>> l[0] = 0
>>> l
[0, 2, 3]
>>> t1
([0, 2, 3], 'a', True)
>>> l1
[[0, 2, 3], [0, 2, 3], 1.0, None]
>>> t1[0][1] = 0
>>> l
[0, 0, 3]
>>> t1
([0, 0, 3], 'a', True)
```

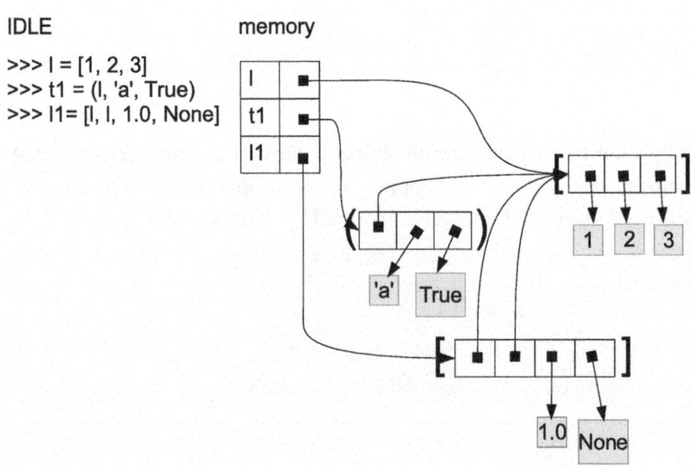

Fig. 7.4 Memory structure for list nested in tuples and lists

```
>>> l1
[[0, 0, 3], [0, 0, 3], 1.0, None]
>>> l1[1][2] = 0
>>> l
[0, 0, 0]
>>> t1
([0, 0, 0], 'a', True)
>>> l1
[[0, 0, 0], [0, 0, 0], 1.0, None]
```

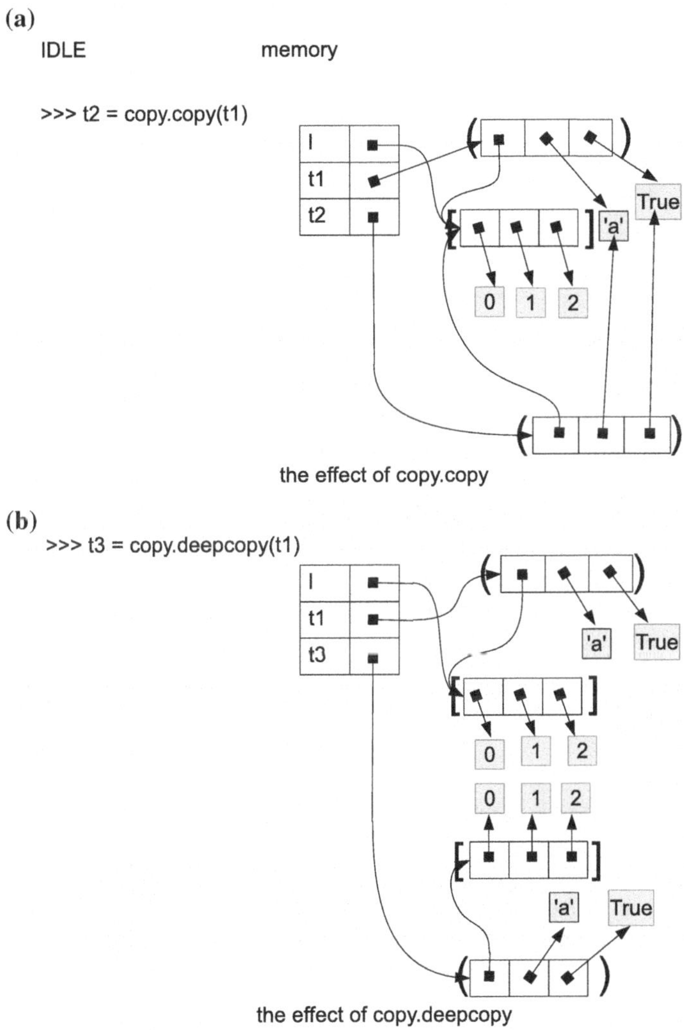

Fig. 7.5 Effect of *copy.copy* and *copy.deepcopy*

Note in the example above that although the *tuple* object *t*1 is immutable, the *list* *t*1[0] can be modified. This is because a *tuple* memory object consists only of the container structure, which is a sequence of binding slots. As shown in Fig. 7.4, the binding links of *t*1 cannot be changed, but the objects bound by the links do not belong to the tuple structure itself. Their mutability is decided by their respective type.

Copying of *t*1 and *l*1 will yield more references to the same *list* object.

```
(continued from above)
>>> import copy
>>> t2 = copy.copy(t1)
>>> l2 = l1[:] # same as copy.copy(l1)
>>> t2[0][1] = 1
>>> l2[0][2] = 2
>>> l
[0, 1, 2]
>>> t1
([0, 1, 2], 'a', True)
>>> t2
([0, 1, 2], 'a', True)
>>> l1
[[0, 1, 2], [0, 1, 2], 1.0, None]
>>> l2
[[0, 1, 2], [0, 1, 2], 1.0, None]
```

In the example above, the five objects *l*, *t*1, *t*2, *l*1 and *l*2 share 7 references to the same *list* object. The memory structures of *l*, *t*1 and *t*2 are illustrated in Fig. 7.5a. As can be seen from the figure, *t*2 contains references to the same objects as *t*1 does. The correlation between *l*2 and *l*1 is the same as that between *t*2 and *t*1.

In order to copy not only the *tuple* object *t*1, but also the objects contained in the *tuple* object, the function *copy.deepcopy* can be used instead of *copy.copy*. *copy.deepcopy* makes a deep copy of a container object by recursively following binding links inside the container. Following the binding links recursively, all reachable objects in the container object are copied. For example,

```
>>> t3 = copy.deepcopy(t1)
>>> t3[0][0] = 3
>>> l
[0, 1, 2]
>>> t1
([0, 1, 2], 'a', True)
>>> t3
([3, 1, 2], 'a', True)
```

When *t*3 is modified, *l* and *t*1 are not affected, because *t*3 contains its own copies of all the sub items recursively, as shown in Fig. 7.5b.[2] Deep copying of *tuples*, *lists* and other container objects follows the same rules.

[2]Due to the caching of small numbers introduced in Chap. 2, the integers 1, 2, 3, the string 'a' and the object *None* are not copied. The figure is for illustration of a simple recursive copying mechanism.

7.2.3 Lists and Loops

Item iteration. Similar to *tuples*, *list* objects are *iterable*, they can be iterated through by using a *for* loop.

```
>>> l1 = ['a', lambda x:x+1, None]
>>> for x in l1:
...     print x, type(x)
...
a <type 'str'>
<function <lambda> at 0x1021d6f50> <type 'function'>
None <type 'NoneType'>
>>> l2 = [1, 2, 3, 4, 5, 6, 7, 8, 9]
>>> for x in l2:
...     print x
...
1
2
3
4
5
6
7
8
9
```

The first *for* loop above prints a *list* of objects and their types, while the second *for* loop above prints the range of integers from 1 to 9, as indicated by the *list*. Iteration through a sequence of numbers is frequently needed in programming, and Python provides a built-in function, *range*, which returns simple number sequences as *list* objects to facilitate number iteration. The *range* function can take one, two or three arguments. When a single argument n is given, the *range* function returns the *list* [0, 1, ..., $n-1$]. If $n \leq 0$, an empty *list* is returned. When two arguments s and e are given, the *range* function returns the *list* [$s, s+1, s+2, ..., e-1$]. If $s \geq e$, an empty *list* is returned. When three arguments, s, e and *step* are given, the *range* function returns the *list [$s, s+step, s+2 \cdot step ..., e'$]*. If $step > 0$, e' is the largest integer in the sequence that is smaller than e. If $step < 0$, e' is the smallest integer in the sequence that is larger than e. If *step=0*, an error is raised.

```
>>> range(5)
[0, 1, 2, 3, 4]
>>> range(20) # making a long sequence
[0, 1, 2, 3, 4, 5, 6, 7, 8, 9, 10, 11, 12, 13, 14, 15,
  16, 17, 18, 19]
>>> range(-10) # results in an empty sequence
[]
>>> range(1, 6)
[1, 2, 3, 4, 5]
>>> range(2, 30, 7)
[2, 9, 16, 23]
>>> range(1, -9, -2)
[1, -1, -3, -5, -7]
>>> range(1, 5, 0)
```

```
Traceback (most recent call last):
  File "<stdin>", line 1, in <module>
ValueError: range() step argument must not be zero
```

The *range* function is frequently used with the *for* loop. A *for* loop with a range function can provide a simple and succinct way of repeating the same task multiple times.

```
>>> for i in range(3):
...     print "hello"
...
```

The program above prints the message "hello" 3 times, without making use of the loop variable *i*.

The *range* function offers a new way to iterate through all the items in an iterable object. Different from using a *for* statement to iterate through an iterable object directly, the *range* function allows the use of a *for* statement to iterate through all the indices from 0 to the length of a sequence object.

```
>>> l = [1, 5, 4, 2, 3]
>>> for x in l: # direct iteration
...     print x
...
1
5
4
2
3
>>> for i in range(len(l)): # index iteration
...     print l[i]
...
1
5
4
2
3
```

Index iteration allows flexible orders of iteration. For example, the code below shows two ways to print the items of a *list* in the reverse order without modifying the *list*, both by index iteration.

```
>>> l = [1, 5, 4, 2, 3]
>>> for i in range(len(l)):
...     print l[len(l)-i-1]
...
3
2
4
5
1
>>> for i in range(len(l)-1, -1, -1): # [4, 3, 2, 1, 0]
...     print l[i]
...
3
2
```

```
4
5
1
```

The example below iterates through every second item in a list:

```
>>> l = [1, 5, 4, 2, 3]
>>> for i in range(0, len(l), 2):
...     print l[i]
...
1
4
3
```

Index iteration also allows iterating through a half of a sequence object. For example, consider the palindrome number problem again. Index iteration allows the problem to be solved without converting the number into a *list* and making of copy of the *list*. The solution is based on the observation that if the ith digit on the left is the same as the ith digit on the right for all i, the corresponding number will be the same after the digits are reversed. Based on symmetry, as long as the left half of a *list* satisfies the condition above, the whole *list* satisfies the condition. The idea can be implemented as searching for a digit that is different from its counterpart. If such a digit is found, then the number is not a palindrome. Otherwise it is a palindrome.

```
>>> def palindrome(n):
...     s = str(n)
...     for i in range(len(s)/2):
...         if s[i] != s[len(s)-i-1]: # mirror digits
...             return False
...     return True
...
>>> palindrome(12321)
True
>>> palindrome(1111)
True
>>> palindrome(123)
False
```

Enumeration. Python provides another built-in function for sequence iteration: the *enumerate* function takes a sequence argument and returns the enumeration of pairs of indices and the corresponding items in the sequence object. The *enumerate* function is used with *for* loops, providing a third way of iterating through an iterable object, by enumerating indices and items simultaneously.

```
>>> l = ['a', 'b', 'c']
>>> for (i, x) in enumerate(l):
...     print 'The {}th object in l in {}'.format(i+1,
    x)
...
The 1th object in l in a
The 2th object in l in b
The 3th object in l in c
```

The *for* loop above makes a *tuple* assignment at each iteration, assigning the current item in *enumerate(l)* to the tuple (i, x). As introduced in Chap. 5, the *tuple* assignment results in the variable *i* taking the value of the current index, and the variable *x* taking the value of the current item in *l*.

For an example of using the *enumerate* function, consider again the problem of finding the maximum value from a sequence of numbers, which was discussed in Chap. 5. Suppose that the sequence of numbers is represented by a *tuple*, and the index of the maximum value must be returned in addition to the maximum value itself. The function *argmax* below solves this problem.

```
>>> def argmax(t):
...     mx = float('-inf')
...     mi = 0
...     for i, x in enumerate(t):
...         if x > mx:
...             mx = x
...             mi = i
...     return mx, mi
...
>>> argmax([1,5,3])
5,1
```

The function *argmax* above takes a *tuple* argument and returns a *tuple* containing the maximum value and its index. It works by iterating through the input *tuple*, keeping the maximum value so far by mx, and the corresponding index of the maximum value by mi at each iteration. mx and mi are updated simultaneously.

Note from the example above that similar to the assigned statement, in a *return* statement, the brackets of a tuple can be omitted.

In summary, there are three ways to iterate through a sequence object using a *for* loop, including to enumerate the items directly (item iteration), to enumerate the indices using the *range* function (index iteration), and to enumerate both the indices and the items using the *enumerate* function.

One further note on lists and loops is that, when a *list* is iterated through by using a *for* loop, the content of the *list* can be modified. This can change the behavior of the loop, and it is recommended to avoid such modifications. For example,

```
>>> def avoid():
...     l = range(10)
...     for x in l:
...         print x
...         l.pop(-1)
...
>>> avoid()
0
1
2
3
4
```

The program above prints 5 numbers only, even though the original *list* object *l* contains 10 numbers. This is because in each loop iteration, the last item of *l* is removed. The value of *l* after the execution is [0, 1, 2, 3, 4].

7.2.4 Lists and Function Arguments

Like other objects, *list* objects can be passed as arguments to functions, or taken as return values. In the example below, the function *duplicate* takes a *list* argument and modifies it by duplicating its content.

```
>>> def duplicate(l):
...     l.extend(l)
...
>>> l1 = [1, 2, 3]
>>> duplicate(l1)
>>> l1
[1, 2, 3, 1, 2, 3]
```

In the example above, the global *list* $l1$ is passed as an argument to the function *duplicate*. When the function call is evaluated, the local argument l is assigned the value of $l1$. As a result, l and $l1$ are bound to the same *list* object. When l is modified by calling the *extend* method of l, $l1$ is modified simultaneously.

The program below, however, does not change the global variable $l1$.

```
>>> def duplicate(l):
...     l = l+l
...
>>> l1 = [1, 2, 3]
>>> duplicate(l1)
>>> l1
[1, 2, 3]
```

The example above contains a common programming mistake when working with *lists*. The reason that the global vairable $l1$ is not modified is that, the assignment statement $l = l+l$ does not modify the object $l1$. When the function call is evaluated, the local parameter l is first assigned the value of $l1$, and bound to the same object. However, when the statement $l = l + l$ is executed, the local identifier l is assigned to a new object, which is the concatenated list $l + l = l1 + l1$. After the assignment statement is executed, the local variable l and the global variable $l1$ are bound to two different objects, without $l1$ being modified. It is worth remembering that the only ways to modify a *list* object include the *setitem*, *setslice*, *delitem* and *delslice* operations, and calls to *list* modification methods such as *append*, *extend* and *remove*. The assignment statement changes the binding table but does *not* modify objects.

List parameters. Python allows the definition of a function that can take an arbitrary number of optional arguments. This is convenient when one does not know how many argument will be specified in a function call, but needs the function to handle a variable number of arguments. This mechanism is allowed by defining a special *list parameter* with a preceding * symbol. For example, consider extending the function *sum* in Chap. 6, which takes two or three arguments and returns their sum, into a function that takes two or more arguments and returns their sum.

```
>>> def sum(x, y, *z):
...     s = x+y              # first two argument
...     for i in z:          # third argument on words
...         s += i
```

```
...     return s
...
>>> sum(1, 2)
3
>>> sum(1, 2, 3)
6
>>> sum(1, 2, 5, 6, 7)
21
```

The function *sum* above takes two or more arguments, putting any argument beyond the second argument into a *list*, and assigning the *list* to the parameter z. The function body performs standard summation, with the sum s being initialized to the value $x + y$, and updated incrementally with each element in z.

At most one special list parameter can be put into the definition of a function, and it must be put after all normal parameters (including those with default valves). In a function call, all the normal and default arguments are assigned their values first, with any additional arguments specified in the function call being put into a *list* in their order, and then assigned to the special *list* argument. A special *list* argument can be passed on from one function to another. For example:

```
>>> def f(x,*y):
    return x+g(*y)
>>> def g(*x):
    print len(x), 'arguments passed!'
    return x[0]
>>> f(1,2,3)
2 arguments passed!
3
>>> f(1,2)
1 arguments passed!
3
```

In the example above, f calls g with *y, passing all the extra arguments to g. In g, these arguments are put into the *list* x, and the first is returned. In fact, any *list* can be passed as arguments in function calls. By adding a preceding *, a *list* is expanded to fill individual arguments. For example:

```
>>> def f(x,y):
        return x+y
>>> l=[1,2]
>>> f(*l)
3
```

In the example above, the *list* l is expanded in the function call $f(*l)$, with its two elements being assigned to x and y, respectively.

7.2.5 Lists and Function Return Values

A *list* object can be constructed in the execution of the function body during a function call, and taken as the return value. For example, consider a function that takes a *list* of numbers as the only argument, and returns a *list* containing all the positive numbers in the *list*. The function can be written using the idea of generalized summation, with the output s being initialized as an empty *list*, and incrementally extended by iterating over all the items of the argument.

```
>>> def findpositive(l):
...     s = []
...     for x in l:
...         if x>0:
...             s.append(x)
...     return s
...
>>> l1 = findpositive([1, -1, 2, -3, 5, -7])
>>> l2 = findpositive([1, 2, 3])
>>> l3 = findpositive([-1])
>>> l1
[1, 2, 5]
>>> l2
[1, 2, 3]
>>> l3
[]
```

In the program above, the function *findpositive* iterates through the items of l. At each iteration, if the current item is positive, it is appended to the output s; otherwise no action is taken.

As introduced in the previous chapter, in memory, each call to *findpositive* constructs a new local scope, with its own binding table. When the function body is executed, a new *list* object is created by the first statement $s = []$, and bound to the local identifier s. s is updated through the execution of the *for* statement in the function body, and taken as the return value of the function call.

The three calls to *findpositive* in the code return three different *list* objects, which are bound to the global identifiers $l1$, $l2$ and $l3$, respectively. After the function calls, the local binding tables of the function calls will be removed by the garbage collector. However, the *list* objects that are returned by the function calls will not be removed, because they are still bound to the global identifiers $l1$, $l2$ and $l3$.

As another example of *lists* and return values, the function *negate* below takes a *list* argument and returns a *list* object that contains the negations of the items of the argument. It works by making a copy of the list argument, and then iterating through the copy and negating each value.

```
>>> def negate(l):
...     s = l[:]
...     for i in range(len(s)):
...         s[i] = -s[i]
...     return s
...
>>> negate([1, -2, 3, -5, 0])
[-1, 2, -3, 5, 0]
```

There are two things to note about the *negate* function above. First, the *for* loop must iterate through the indices of s instead of the items of s directly, because a modification to s requires the use of item indices to specify the items to be changed.

Second, the method behind *findpositive*, which initializes an empty return value *list*, and then incrementally adds items to the *list*, can be used for *negate* also. The *negate* function above takes a different approach, which initializes the *list* return value as a copy of the *list* argument, and incrementally modifies it. This is also a common way of thinking when working with *lists*.

7.2.6 Initializing a List

There are four common ways to **initialize a list**. The first is to make an empty *list* (i.e. []); the second is to make a *list* of length N, consisting of the same constant x, for which the expression $[x] * N$ can be used; the third is to make a *list* that contains a number sequence, for which the *range* function can be used; the fourth is to make a copy of a given sequence, for which a copying operation can be used. If the *list* to be initialized is more complex then the cases above, it should be initialized using a specific loop.

On initializing a *list* of constants, there is one common mistake to avoid. If the constant item is mutable, then the *list* should *not* be initialized using the $[x] * N$ expression, because in that way all the items in the *list* are bound to the same mutable object. For example,

```
>>> l = [[]]*3
>>> l
[[], [], []]
>>> l[0].append(1)
>>> l
[[1], [1], [1]]
```

As shown by the example above, when one item in the *list* is changed, all the other items are affected. To avoid such a mistake, one can use the expression $[x] * N$ only with immutable constants x.

A list of empty lists can be initialized using a simple loop.

```
>>> l = []
>>> for i in range(3): # repeat 3 times
...     l.append([]) # every time [] makes a new list
...
>>> l
[[], [], []]
>>> l[0].append(1)
>>> l
[[1], [], []]
```

When initializing a *list* by making a copy of a sequence object, the given sequence object can be a *list*, a tuple or other sequence types. In the case of a *list*, the *list* copying techniques introduced earlier in this chapter can be used. In the case of a *tuple*, the *list* function converts a *tuple* into a *list* by making a *shallow* copy of the *tuple*.

```
>>> t = (1, 2, 3)
>>> l = list(t) # initialization from tuple
>>> l
```

```
[1, 2, 3]
>>> s ="123" # initialization from string
>>> l = list(s)
>>> l
['1', '2', '3']
```

Sometimes, however, the input sequence contains mutable items, and it can be necessary to use *copy.deeepcopy* to copy the sequence recursively.

7.2.7 Lists and Sequential Data Structures

Data structure is an important concept in programming; it refers to the organization of complex data. *Tuples* and *lists* by themselves are data structures, representing a sequential way of organizing data, with items being accessed by their indices. There are other common types of sequential data structures that can be represented by *lists*.

The **stack** is a sequential data structure that allows only two types of operations:

- PUSH, which adds an item onto the top of a stack;
- POP, which removes the item on the top of a stack, returning its value.

The stack is a commonly used data structure in computer science. It is special in is that items must be processed in a last-in-first-out (LIFO) order. The stack can be used to model a pile of books in a container, and a sequence of nested function calls, such as the one shown in Table 6.1 in Chap. 6.

Stacks can be represented by a list in Python. The program *stack_fun.py* below contains two functions, *push* and *pop*. The *push* function takes two arguments, representing a stack and an item, respectively, pushing the item onto the stack. The return value of the function is *None*. The *pop* function takes a stack argument and pops the top of the stack, returning its value.

```
[stack_fun.py]
def push(l, s):
   l.append(s)

def pop(l):
   return l.pop(-1)
```

The program below demonstrates the LIFO behavior of a stack.

```
>>> from stack_fun import *
>>> s = []          # BOTTOM=TOP
>>> push(s, 0)      # BOTTOM, 0=TOP
>>> push(s, 1)      # BOTTOM, 0, 1=TOP
>>> push(s, 2)      # BOTTOM, 0, 1, 2=TOP
>>> print pop(s) # BOTTOM, 0, 1=TOP
2
>>> print pop(s) # BOTTOM, 0=TOP
1
>>> push(s, 3)      # BOTTOM, 0, 3=TOP
>>> print pop(s) # BOTTOM, 0=TOP
```

```
3
>>> push(s, 4)    # BOTTOM, 0, 4=TOP
>>> push(s, 5)    # BOTTOM, 0, 4, 5=TOP
>>> print pop(s) # BOTTOM, 0, 4=TOP
5
>>> print pop(s) # BOTTOM, 0=TOP
4
>>> print pop(s) # BOTTOM=TOP
0
```

A **queue** is a sequential structure that allows only two types of operations:

- ENQUEUE, which adds an item to the back of a queue;
- DEQUEUE, which removes the item in the front of a queue, returning its value.

In contrast to a stack, items of a queue must be processed in a first-in-first-out (FIFO) order. The queue can be used to model a line for movie tickets, and a sequence of instructions to be executed in a Python program.

Similar to stacks, queues can be represented by a *list* in Python. The *queue_fun.py* below contains two functions, *enqueue* and *dequeue*. The *enqueue* function takes two arguments representing a queue and an item, respectively, pushing the item into the queue. The return value of the function is *None*. The *dequeue* function takes a queue argument and pops the front of the queue, returning its value.

```
def enqueue(l, s):
    l.append(s)

def dequeue(l):
    return l.pop(0)
```

The program below demonstrates the FIFO behavior of a queue.

```
>>> from queue_fun import *
>>> q = []              # FRONT=BACK
>>> enqueue(q, 0)       # FRONT=0, BACK
>>> enqueue(q, 1)       # FRONT=0, 1, BACK
>>> enqueue(q, 2)       # FRONT=0, 1, 2, BACK
>>> print dequeue(q) # FRONT=1, 2, BACK
0
>>> print dequeue(q) # FRONT=2, BACK
1
>>> enqueue(q, 3)       # FRONT=2, 3, BACK
>>> print dequeue(q) # FRONT=3, BACK
2
>>> enqueue(q, 4)       # FRONT=3, 4, BACK
>>> enqueue(q, 5)       # FRONT=3, 4, 5, BACK
>>> print dequeue(q) # FRONT=4, 5, BACK
3
>>> print dequeue(q) # FRONT=5, BACK
4
>>> print dequeue(q) # FRONT=BACK
5
```

The **priority queue** is a sequential data structure that allows only two operations, ENQUEUE and DEQUEUE, with the ENQUEUE operation inserting an item into a priority queue, and the DEQUEUE operation removing the item with the highest priority from the priority queue, returning its value. Different from both stacks and queues, for which the removing order is dependent on the inserting order, priority queues determine the removing order solely based on the priority of items. Items are processed in the highest-priority-first-out order.

A priority queue can be represented by a *list*, organized in a **heap** structure. Here a heap does not order all items by their priority, but organizes items in such a way that the item with the highest priority is always in the front, and it is highly efficient to find the item with the next highest priority when the current front of the priority queue is removed.

Python provides a module, *heapq*, which contains two operations that correspond to the ENQUEUE and DEQUEUE operations for a priority queue, respectively:

- *heappush(h,x)*, which takes two arguments representing a heap and an item, respectively, pushing the item onto the heap. The function returns *None*.
- *heappop(h)*, which takes a heap argument, removes the front item of the heap, and returns its value.

The *heapq* module treats a numerically smaller item as having a higher priority. The code below demonstrates the highest-priority-first-out behavior of a priority queue.

```
>>> import heapq
>>> h = []
>>> heapq.heappush(h, 1)      # highest priority=1
>>> heapq.heappush(h, 5)      # highest priority=1
>>> heapq.heappush(h, 2)      # highest priority=1
>>> print heapq.heappop(h)    # 1 popped
1
>>> heapq.heappush(h, 3)      # highest priority=2
>>> heapq.heappush(h, 4)      # highest priority=2
>>> print heapq.heappop(h)    # 2 pop
2
>>> print heapq.heappop(h)    # 3 pop
3
>>> print heapq.heappop(h)    # 4 pop
4
>>> heapq.heappush(h, 0)      # highest priority=0
>>> print heapq.heappop(h)    # 0 pop
0
>>> print heapq.heappop(h)    # 5 pop
5
```

Exercises

1. What is the value of *l* after the following programs are executed?

   (a) List in tuple

   ```
   l = [1, 2, 3]
   t = [1, 'a']
   t[0] = 0
   ```

(b) List in function

```
def f(l):
    l[0] = 0
l = [1, 2, 3]
f(l)
```

(c) List and function

```
def f():
    l[0] = 0
l = [1, 2, 3]
f()
```

(d) List in function

```
def f(l):
    l = [2, 3, 4]
l = [1, 2, 3]
f(l)
```

(e) What are the values of *x* and *y* after the following code is executed?

```
def myappend(l):
    l.append(1)
    return l
x = [1, 2]
y = myappend(x)
y.append(3)
```

2. Which of the following programs are correct?

(a) Sum up an arbitrary number of numbers.

```
def sum(x, y=0, *z):
    s = x+y
    for i in z:
        s += i
    return s
```

(b) Sum up an arbitrary number of numbers.

```
def sum(x, *y, *z):
    s = x
    for i in y+z:
        s += i
    return s
```

(c) Sum up an arbitrary number of numbers.

```
def sum(x, *y, z=0):
    s = x+z
    for i in y:
        s += i
    return s
```

(d) Append the number 1 to the *list l*.

```
l = l.append(1)
```

(e) Return the *list* [0, 1, 2, 3, 4].

```
def f():
    for i in range(5):
        return i
```

3. Write a function *f*, which takes a function *g* as the first argument, and an arbitrary number of additional arguments, returning the return value of a function call to *g* with the other arguments to *f* being the arguments. For example,

```
>>> f(math.sqrt, 49)
7.0
>>> f(lambda x, y, z: x + y + z, 1, 2, 3)
6
>>> f(math.log, math.e)
1.0
>>> f(math.log, 100, 10)
2.0
```

Compare the function with the answer to question 3 in Chap. 6.

4. Write a function that takes a *list* of numbers as the only argument and returns a *list* of numbers that contains

   (a) all the multiples of 3 in the argument;
   (b) the second power of each item in the argument;
   (c) the items in the arguments, with only one instance from each consecutive run of the same number. (e.g. [1,2,2,3,3,3,4], [1,2,3,4])

5. Write a function that takes a *list* as the only argument and returns

   (a) the number of odd items in the list;
   (b) the average of all the items in the list;
   (c) the sum of squared items in the list.

6. The built-in function *zip* takes two *list* arguments, and returns a *list* that contains zipped pairs of corresponding elements in the input arguments.

```
>>> zip([1, 2, 3],['A', 'C', 'E'])
[(1, 'A'), (2, 'C'), (3, 'E')]
```

Write a function, *unzip*, that takes a *list* of pairs and performs the reverse function to *zip*.

Chapter 8
Sequences, Mappings and Sets

Several sequential data structures have been introduced in this book, including strings, *tuples*, *lists*, and structures that can be implemented using *lists*, such as stacks, queues and priority queues. This chapter further gives a set of commonly used methods for sequential types, before moving on to two typical container data structures that are not sequential: *dicts* and *sets*. While *dicts* represent unordered mappings from keys to values, a *set* represents an unordered container of immutable objects. Both *dicts* and *sets* are mutable. This chapter finishes by introducing a set of bitwise operators, which enable the implementation of *sets* using non-negative integers.

8.1 Methods of Sequential Types

The previous chapter has introduced a set of methods for *list* modification. In addition to these methods, there is another set of methods that return information about a *list* object without modifying it. List information methods typically apply to immutable sequential objects, such as strings and *tuples*, also. This section introduces common methods for sequential types, and in particular for strings, by categorizing them according to the functionality.

One commonly-used method is *index*, which takes one argument and returns the index of the first occurrence of the argument in the sequence. If the argument is not in the sequence, a *value error* is reported.

```
>>> l = [1, 2, 3]
>>> l.index(2)
1
>>> l.index(4)
Traceback (most recent call last):
  File "<stdin>", line 1, in <module>
ValueError: 4 is not in list
>>> t = (1.0, 2.5, 3.3, 2.5)
>>> t.index(2.5)
```


Y. Zhang, *An Introduction to Python and Computer Programming*,
Lecture Notes in Electrical Engineering 353, DOI 10.1007/978-981-287-609-6_8

```
1
>>> t.index(3.0)
Traceback (most recent call last):
  File "<stdin>", line 1, in <module>
ValueError: tuple.index(x): x not in tuple
>>> s = "abcdefgfedcba"
>>> s.index('c')
2
>>> s.index('cba')
10
>>> s.index('h')
Traceback (most recent call last):
  File "<stdin>", line 1, in <module>
ValueError: substring not found
```

As shown in the example above, the *index* method for strings can also take a string argument that consists of more than one characters, returning the index of the first character in the string if the argument is a sub string.

Another useful sequence information method is *count*, which takes one argument and returns the number of occurrences of the argument in the sequence. If the argument does not occur in the sequence, 0 is returned.

```
>>> t = ('2', '3', '5', '7', '3', '5', '6', '3', '2')
>>> t.count('3')
3
>>> t.count(3)
0
>>> t.count('1')
0
>>> s = 'exciting'
>>> s.count('i')
2
>>> s.count('it')
1
```

As shown by the example above, the string method *count* can also take a string argument that consists of more than one characters.

As a sequence type for I/O, the string type has a number of useful methods that the *tuple* and *list* types do not have. String methods can be classified into two basic categories. The first category returns information about a string, and the second category formats a string. Because string objects are immutable, string formatting methods return a copy of the string object, rather than modifying the object in place.

A set of commonly-used string information methods include:

- *isupper()*, which takes no arguments and returns *True* if there is at least one cased character in the string and all cased characters in the string are in upper case, and *False* otherwise;
- *islower()*, which takes no arguments and returns *True* if there is at least one cased character in the string and all cased characters in the string are in lower case, and *False* otherwise;
- *isalpha()*, which takes no arguments, and returns *True* if the string is non-empty and all the characters in the string are alphabetic, and *False* otherwise;

- *isdigit()*, which takes no arguments, and returns *True* if the string is non-empty and all the characters in the string are digits, and *False* otherwise;
- *isspace()*, which takes non arguments, and returns *True* if the string is non-empty and all the characters in the string are whitespaces (e.g. spaces, tabs and new lines), and *False* otherwise.

```
>>> s1 = 'abcdef'
>>> s2 = 'A1'
>>> s3 = 'ABC'
>>> s4 = '__name__'
>>> s5 = 'hello!'
>>> s6 = 'name@institute'
>>> s7 = '12345'
>>> s8 = '123 456'
>>> s9 = ''
>>> s10 = ''
>>> s1.isupper()
False
>>> s2.isupper()
True
>>> s3.isupper()
True
>>> s8.isupper()
False
>>> s9.isupper()
False
>>> s10.isupper()
False
>>> s1.islower()
True
>>> s4.islower()
True
>>> s5.islower()
True
>>> s1.isalpha()
True
>>> s2.isalpha()
False
>>> s7.isalpha()
False
>>> s8.isalpha()
False
>>> s10.isalpha()
False
>>> s1.isdigit()
False
>>> s4.isdigit()
False
>>> s6.isdigit()
False
>>> s7.isdigit()
True
>>> s8.isdigit()
False
>>> s9.isdigit()
```

```
False
>>> s8.isspace()
False
>>> s9.isspace()
True
>>> s10.isspace()
False
```

A set of string information methods that search for a sub string includes:

- *rindex(s)*, which takes a string argument *s* and returns the last occurrence of the argument in the string. If the string argument contains more than one characters, the index of the first character is returned. If the string argument is not a sub string of the string, a value error is reported. *rindex* is the counterpart of *index* for string objects, returning the first occurrence of the sub string from the right, rather than from the left.
- *find(s)*, which takes a string argument *s* and returns the first occurrence of the argument in the string. If the string argument contains more than one characters, the index of the first character is returned. If the string argument is not a sub string of the string, *find* returns –1. *find* can be treated as an alternative to *index* for string objects, with the difference being the responce when the string argument is not a sub string: while the former returns –1, the latter raises a value error.
- *rfind(s)*, which takes a string argument *s* and returns the last occurrence of the argument in the string. If the string argument contains more than one characters, the index of the first character is returned. If the string argument is not a sub string of the string, *rfind* returns –1. *rfind* is the counterpart of *find*, returning the first occurrence of the string argument from the right, rather than from the left.

```
>>> s = 'abcdefdefabc'
>>> s.index('abc')
0
>>> s.find('def')
3
>>> s.rfind('de')
6
>>> s.rindex('a')
9
>>> s.index('abd')
Traceback (most recent call last):
  File "<stdin>", line 1, in <module>
ValueError: substring not found
>>> s.rfind('abd')
-1
```

The *index*, *rindex*, *find* and *rfind* methods can take optional arguments that specify a slice of the string in which the sub string is looked for. This is achieved by two optional arguments, indicating the start index and the end index of the slice, respectively. The locations of the slicing indices are the same as those by the *getslice* operations. In case only the start index is specified, the slice ends at the end of the string.

```
>>> s = 'abcabcabcdefdefabc'
>>> s.index('abc', 3) # search s[3:]
3
>>> s.find('abc', 5, -2) # search s[5:-2]
6
```

A set of convenient methods for checking the beginning and end of a string include:

- *startswith(s)*, which takes a string argument *s* and returns *True* if the string starts with the argument, and *False* otherwise;
- *endswith(s)*, which takes a string argument *s* and returns *True* if the string ends with the argument, and *False* otherwise.

```
>>> s = 'abcdefghi'
>>> s.startswith('a')
True
>>> s.startswith('abc')
True
>>> s.endswith('d')
False
>>> s.endswith('hi')
True
```

The methods *startswith* and *endswith* can also take a tuple of strings as the argument, in which case the return value is *True* if the string starts with any string in the tuple, and *False* otherwise.

```
>>> s = 'abcdefghi'
>>> s.startswith(('abc', 'def'))
True
>>> s.startswith(('a', 'b', 'c'))
True
>>> s.endswith(('def', 'abc'))
False
>>> s.endswith(('ghi', 'hi', 'i'))
True
```

A set of string formatting methods that modify the cases of cased characters includes:

- *upper()*, which takes no arguments and returns a copy of the string object with all cased characters converted to the upper case;
- *lower()*, which takes no arguments and returns a copy of the string object with all cased characters converted to the lower case;
- *swapcase()*, which takes no arguments and returns a copy of the string object with all upper-case characters converted into the lower case, and vice versa.

```
>>> s = 'Abc!'
>>> s.upper()
'ABC!'
>>> s.lower()
'abc!'
>>> s.swapcase()
'aBC!'
```

A set of string formatting methods that adds or removes whitespaces on the ends of a string includes:

- *ljust(x)*, which takes an input argument *s* that specifies a length, and returns a copy of the string left justified in a string of the specified length. Spaces are padded on the right if the length is longer than the length of the string, while no truncation is performed if the length is smaller than the length of the string.
- *rjust(x)*, which takes an input argument *s* that specifies a length, and returns a copy of the string right justified in a string of the specified length. Spaces are padded on the left if the length is larger than the length of the string, while no truncation is performed if the length is smaller than the length of the string.
- *lstrip()*, which takes no arguments and returns a copy of the string with whitespaces on the left being stripped.
- *rstrip()*, which takes no arguments and returns a copy of the string with whitespaces on the right being stripped.
- *strip()*, which takes no argument and returns a copy of the string with whitespaces on both sides being stripped.

```
>>> s = 'abc'
>>> s.ljust(5)
'abc  '
>>> s.rjust(6)
'   abc'
>>> s.ljust(2)
'abc'
>>> s = ' abc\n\t'
>>> s
' abc\n\t'
>>> s.lstrip()
'abc\n\t'
>>> s.rstrip()
'abc'
>>> s.strip()
'abc'
```

The methods above can be generalized to insert or strip arbitrary characters on the ends of a string. In particular, *ljust* and *rjust* can take an additional argument that specifies a padding character, while *lstrip*, *rstrip* and *strip* can take an additional argument that specifies a set of characters to be stripped.

```
>>> s = '123'
>>> s.ljust(5, '0') # pad '0'
'12300'
>>> s.rjust(6, '-') # pad '-'
'---123'
>>> s = 'aabcdceft'
>>> s.lstrip('a') # strip 'a'
'bcdceft'
>>> s.lstrip('bac') # the set {'b', 'a', 'c'}
'dceft'
```

```
>>> s.rstrip('ag') # the set {'a', 'g'}
'aabcdceft'
>>> s.strip('gabf') # {'g', 'a', 'b', 'f'}
'cdceft'
```

A method that splits a string into a *list* of sub strings is *split*, which takes a single argument specifying the deliminator, and returns the *list* that results from the splitting.

```
>> s = '1 2 3'
>>> s.split(' ')
['1', '2', '3']
>>> s = '1<2<3'
>>> s.split('<')
['1', '2', '3']
>>> s = '1==2==3'
>>> s.split('=')
['1', '', '2', '', '3']
>>> s.split('==')
['1', '2', '3']
>>> s = '1!=2!=3'
>>> s.split('!=')
['1', '2', '3']
```

As shown by the example above, an empty string results in the splitting of two consecutive deliminators, and a deliminator can be an arbitrary string that contains one or more characters.

The input deliminator is optional; when it is not specified, all consecutive runs of whitespace characters are taken as deliminators.

```
>>> s = ' 1  2\t\t3\n4'
>>> s
' 1  2\t\t3\n4'
>>> s.split()
['1', '2', '3', '4']
```

The reverse method to *split* is *join*, which takes a list argument and joins the list into a string with the string being the concatenator.

```
>>> s = ' '
>>> s.join(['1', '2', '3'])
'1 2 3'
>>> l = ['abc', 'd', 'ef']
>>> '--'.join(l)
'abc--d--ef'
>>> '-'.join([1, 2, 3])
Traceback (most recent call last):
  File "<stdin>", line 1, in <module>
TypeError: sequence item 0: expected string, int found
```

As shown by the examples above, the *split* method returns a *list* of strings, and a *list* of strings should be used as the argument to *join*. Any other *list* items are not allowed by the *join* method, and no implicit conversion is performed.

Using *split* and *join*, one can replace all occurrences of a sub string in a string with another string. For example, the following code replaces all occurrences of 'abc' in a string into '—':

```
>>> s1 = 'abc'
>>> s2 = '---'
>>> s = 'abcdefabcghiabc'
>>> s2.join(s.split(s1))
'---def---ghi---'
```

Python provides a method, *repalce*, for sub string replacement directly. It takes two string arguments, specifying the substring to be replaced and the replacement string, respectively, and returns a copy of the string after replacement.

```
>>> s = 'abcdefabcghiabc'
>>> s.replace('abc', '---')
'---def---ghi---'
```

replace allows an additional argument that specifies a count, so that the first occurrences of the sub string up to the count are replaced.

```
>>> s = 'abcdefabcghiabc'
>>> s.replace('abc', '---', 1)
'---defabcghiabc'
>>> s.replace('abc', '---', 2)
'---def---ghiabc'
```

A final string formatting method is *format*, which is an alternative to the string formatting expression introduced in Chap. 3. The method is bound to a pattern string, and takes arguments that fill pattern fields in the string. A pattern field is formed by a pair of curly brackets, and contains either a number specifying the index of the corresponding argument, or a keyword to be filled by a keyword argument. For example,

```
>>> '{0} + {1} = {2}'.format(1, 2, 1+2)
'1 + 2 = 3'
>>> 'Hello, {0}'.format('Python')
'Hello, Python'
>>> '{2}-{0}-{1}'.format('abc', 'def', 'ghi')
'ghi-abc-def'
>>> '{x}-{y}-{0}'.format('ghi', x='abc', y='def')
'abc-def-ghi'
```

In the example above, the last method call contains two keyword arguments, *x* and *y*, which fill the keyword fields {*x*} and {*y*}, respectively. Note that similar to other function calls, keyword arguments must be placed after non-keyword arguments.

If the arguments are sequentially filled, the indics in the pattern fields can be omitted.

```
>>> '{} + {} = {}'.format(1, 2, 3)
'1 + 2 = 3'
```

Formatting specifications can be given to each pattern field by using ‘:*<pattern>*’, where *<pattern>* follows the pattern syntax in string formatting expressions. For example,

```
>>> '{0:d} + {1:.2f} = {2}'.format(1, 2.0, 1+2.0)
'1 + 2.00 = 3.0'
>>> s = '{0:d} + {1:.2f} = {x:s}'
>>> s.format(1, 2, x=str(1+2))
'1 + 2.00 = 3'
```

In the example above, ':d', ':.2f' and ':s' are used to specify the format of an integer, a floating point number with 2 digits after the decimal point and a string, respectively. Formatting specifications can be used for keyword and non-keyword arguments.

If there are too few arguments, an error will be raised. If there are too many arguments, the first ones will be used to fill the pattern string.

```
>>> s='{} + {} = {}'
>>> s.format(1)
Traceback (most recent call last):
  File "<stdin>", line 1, in <module>
IndexError: tuple index out of range
>>> s.format(1, 2, 3, 4)
'1 + 2 = 3'
```

There is a built-in function, *format(v,s)*, which takes a value *v* and a pattern string *s* as its input arguments, and return a string by calling *s.format(v)*.

```
>>> format(3.1,'.2f')
'3.10'
```

The methods and functions above are frequently used for sequential types, and this section categories them according to the functionality and return type. There are more methods for sequential types. The Python documentation is a useful reference for looking for a pre-defined function before writing one custom function.

8.2 Dicts—A Mutable Mapping Type

The dict type is a container type. Similar to lists, *dict* objects are mutable. On the other hand, unlike lists and tuples, which are *sequential* containers, *dicts* are *associative* containers, which represent mappings from keys to values. Intuitively, *dict* objects resemble dictionaries, which map words to their meanings. When a word is looked up in a dictionary, it is used as a *key* for finding the corresponding *value*, which is the entry explaining the word. *Dicts* generalize the definitions of keys and values, and are useful for mapping student IDs to their GPAs, looking up the price of a stock given a company name, and other *key-value* associations where keys are discontinuous and distinct.

Formally, a *dict* object consists of a set of (*key*, *value*) pairs, which are *items*. A key must be an immutable object, while a value can be any object. The most important operation for a dict object is *lookup*, which returns the corresponding value given a key. As a result, keys must be distinct in a *dict* object, but values do not have to be distinct.

A **dict literal** consists of a comma-separated list of *key*: *value* pairs, enclosed in a pair of curly brackets.

```
>>> d = {1: 'a', 2: 'b', 3: 'c'}
>>> d
{1: 'a', 2: 'b', 3: 'c'}
>>> type(d)
<type 'dict'>
>>> d = {1.0: 1, True: 1+2j, None: 'abc', (1,2): False}
>>> d
{(1, 2): False, 1.0: (1+2j), None: 'abc'}
>>> type(d)
<type 'dict'>
>>> d = {[1,2]: 'a'}
Traceback (most recent call last):
  File "<stdin>", line 1, in <module>
TypeError: unhashable type: 'list'
```

In the example above, the first *dict* represents a mapping from integers to strings. The second *dict* represents a mapping between miscellaneous objects, where the keys include the floating point number 1.0, the Boolean object True, the *None* object and the *tuple* object (1, 2), and the values include the integer 1, the complex number 1 + 2j, the string 'abc' and the Boolean object *False*. The third *dict* literal leads to an error, because the mutable object [1, 2] is used as a key. The literal of an empty *dict* is {}.

When two duplicated keys are defined in a *dict* literal, the latter overrides the former.

```
>>> d = {'a': 1, 'b': 2, 'a': 3}
>>> d
{'a': 3, 'b': 2}
```

In the example above, *d* contains two identical keys, which are mapped into the integers 1 and 3, respectively. Python discards the key value pair ('a', 1), keeping only ('a', 3) for the key 'a'. Python uses the == operator to decide whether two keys are identical. As a result, two numerical values can be identical even if their types are different.

```
>>> d = {1: 'a', 2: 'b', 1.0: 'c'}
>>> d
{1: 'c', 2: 'b'}
```

In the example above, the keys 1 and 1.0 are treated as identical because the expression 1 == 1.0 is *True*. As immutable objects, floating point numbers can be used as keys to *dicts*. However, due to rounding off errors, Python's representation of floating point numbers can be imprecise. These can also be conflit with integer keys as shown above. As a result, it is recommended to avoid the use of floating point numbers as keys when possible.

Dict operators similar to *tuples* and *lists*, the == and != operators can be used to find the equality between *dict* objects. *Dicts* are unordered collections, the order in which items are specified or added to a *dict* object does not affect the value of the *dict* object.

```
>>> d1 = {1: 'a', 2: 'b', 3: 'c'}
>>> d2 = {3: 'c', 1: 'a', 2: 'b'}
>>> d3 = {}
>>> d1 == d2
True
>>> d1 != d3
True
```

The *lookup* operation is the most commonly used operation for *dict* objects. The syntax is formally identical to that of the *getitem* operation for *tuples* and *lists*, which specify the index of an item by a pair of square brackets. The only difference is that for a *dict* object, a key is used in the place of the index for *tuples* and *lists*.

```
>>> d = {1: 'a', 2: 'b', 3: 'c'}
>>> d[1]
'a'
>>> d[3]
'c'
>>> d[5]
Traceback (most recent call last):
  File "<stdin>", line 1, in <module>
KeyError: 5
```

As shown by the example above, a *key error* is raised if the key to lookup is not in the *dict* object.

The dict method *get* is commonly used as an alternative to *lookup*, which allows default values if the specified key is not in the dict:

```
>>> d = {1: 'a', 2: 'b', 3: 'c'}
>>> d.get(3, 'z')
'c'
>>> d.get(5, 'z')
'z'
```

As shown in the example above, *get* takes two arguments, the first being the key, and the second being the default value. If the key is in the *dict*, the return value is the value of the key in the *dict*. Otherwise the default value is returned by *get*.

Similar to the case of *tuples* and *lists*, the operator *in* and *not in* can be applied to a *dict* object, returning whether an object is in the *dict* or not. When applied to a *dict*, the operators return whether an object is a *key* in the *dict*.

```
>>> d = {1: 'a', 2: 'b', 3: 'c'}
>>> 1 in d
True
>>> 4 in d
False
>>> 'a' in d
False
```

Note that the last expression in the example above is evaluated to *False*, because 'a' is not a key in *d*, despite that it is the value of the key 1 in *d*.

The method call *d.get(k, v)* can be regarded as a succinct version of the following function:

```
def defaultlookup (d, k, v):
    if k in d:
        return d[k]
    else:
        return v
```

The *len* function can be applied to *dict* objects, returning the number of items in the *dict*.

```
>>> len({1: 'a', 2: 'b', 3: 'c'})
3
>>> len({})
0
```

Conversion between dicts and other types. *Dicts* can be converted to strings, Boolean objects, *tuples* and *lists*, but not to numbers. The string conversion of a *dict* object represents the literal form of the *dict* object.

```
>>> a = 1.0
>>> b = True
>>> c = 'a'
>>> d = {a: 1, 2: b, c: 3}
>>> str(d)
"{'a': 3, 1.0: 1, 2: True}"
```

The Boolean conversion of a *dict* is *False* only when the *dict* is empty, and *True* otherwise.

```
>>> bool({1: 'a', 2: 'b', 3: 'c'})
True
>>> bool({1: 1})
True
>>> bool({})
False
```

The *tuple* and *list* conversions of a *dict* object consist of all the keys in the *dict* object.

```
>>> d = {1: 1.0, 2: 9, 3: 8}
>>> tuple(d)
(1, 2, 3)
>>> list(d)
[1, 2, 3]
```

Lists and *tuples* can be converted to *dicts* using the *dict* function, but numbers, strings, or Boolean objects cannot. To be convertable to a *dict*, a *tuple* or *list* must be a sequence (*key*, *value*) pairs.

```
>>> d1 = dict(((3, 'a'), (2, 'b'), (3, 'c')))
>>> d2 = dict([(3, 'a'), (2, 'b'), (3, 'c')])
>>> d1
{2: 'b', 3: 'c'}
>>> d1 == d2
True
```

8.2.1 Dict Modification

There are two ways to change a *dict* object. The first is the *setitem* operation, which sets the value of a key directly via the assignment syntax, and the *delitem* operator, which deletes a key-value pair from a *dict* object. The syntax of the *setitem* operation is *d*[*key*] = *value*, where *d* is a *dict* identifier, key is an immutable object and *value* is an *arbitrary* object. If *key* is not an existing key in *d*, (*key*, *value*) is added as a new item in *d*. Otherwise, *value* is used to replace the old value of *key* in *d*. For example

```
>>> d = {1: 'a', 2: 'b', 3: 'c'}
>>> d[0] = '' # adds a new entry to d
>>> d
{0: '', 1: 'a', 2: 'b', 3: 'c'}
>>> d[1] = 'e' # update an existing entry in d
>>> d
{0: '', 1: 'e', 2: 'b', 3: 'c'}
```

The syntax of the *delitem* operation is

```
del d[key]
```

where *d* is a *dict* identifier and key is an immutable object. If *key* is an existing key in *d*, the corresponding item will be removed from *d*. Otherwise a *key error* is raised.

A commonly used method to change a *dict* object is *update*, which takes a single *dict* argument, using it to update the *dict*. In particular, each (*key*, *value*) pair in the argument is enumerated, and used to update the *dict* in the same way as *setitem*. If *key* is not already in the *dict*, the item (*key*, *value*) is added to the *dict* as a new item. Otherwise, the corresponding value of *key* is updated by *value*. The return value of the *update* method is *None*.

```
>>> d = {1: 'a', 2: 'b', 3: 'c'}
>>> d1 = {0: '', 1: 'e'}
>>> d.update(d1)
>>> d
{0: '', 1: 'e', 2: 'b', 3: 'c'}
```

In the example above, *d1* is used to update *d*. There are two items in *d1*, namely (0, ' ') and (1, 'e'). The key 0 is not in *d* before the update, and therefore (0, ' ') is added to *d*. On the other hand, the key 1 is already in *d*, and therefore the existing item (1, 'a') in *d* is updated with the item (1, 'e').

A method to remove an item from a *dict* is *pop*, which takes a single argument. If the argument is a key in the *dict*, the corresponding item is removed, with is value returned. Otherwise, a *key error* is raised.

```
>>> d = {1: 'a', 2: 'b', 3: 'c'}
>>> d.pop(1)
'a'
>>> d
{2: 'b', 3: 'c'}
>>> d.pop(0)
Traceback (most recent call last):
```

```
  File "<stdin>", line 1, in <module>
KeyError: 0
```

pop can take an optional second argument, which specifies the default return value if the first argument is not a key in the *dict*. When the second argument is given, no *key error* is raised.

```
>>> d = {1: 'a', 2: 'b', 3: 'c'}
>>> d.pop(1, 'd')
'a'
>>> d
{2: 'b', 3: 'c'}
>>> d.pop(0, 'd')
'd'
>>> d
{2: 'b', 3: 'c'}
```

A method that removes all the items from a *dict* object is *clear*, which takes no argument, and returns *None*.

```
>>> d = {1: 'c', 2: 'b', 3: 'c'}
>>> d.clear()
>>> d
{}
```

Similar to the case of *lists*, modifications to a *dict* object shared by multiple identifiers results in a simultaneous change of values of all the identifiers.

```
>>> d = {1: 'a', 2: 'b', 3: 'c'}
>>> d1 = d
>>> t1 = (d1, 4, 'd')
>>> def f(d2):
...     d2.update({0: 'e', 1: 'd'})
...     del d2[3]
...
>>> f(t1[0])
>>> d
{0: 'e', 1: 'd', 2: 'b'}
>>> d1
{0: 'e', 1: 'd', 2: 'b'}
>>> t1
({0: 'e', 1: 'd', 2: 'b'}, 4, 'd')
```

In the example above, the identifiers d, $d1$ and $t1[0]$ are bound to the same *dict* object. When f is called, the local identifier $d2$ is bound to the same object as the argument. As a result, when $f(t1[0])$ is called, all the identifiers above are changed.

In addition to *dict* modification methods, the *dict* type also supports a set of methods that return information of a *dict* object, including *keys*, *values* and *items*, which take no argument and return a *list* of *copies* of the keys, values and items of a *dict* object, respectively.

```
>>> d = {1: 'a', 2: 'b', 3: 'c'}
>>> d.keys()
[1, 2, 3]
>>> d.values()
['a', 'b', 'c']
```

```
>>> d.items()
[(1, 'a'), (2, 'b'), (3, 'c')]
```

Because the return values of the methods are copies of the original keys, values and items, modifications to the return values do not affect the original *dict*.

```
>>> d = {1: 'a', 2: 'b', 3: 'c'}
>>> l = d.keys()
>>> l
[1, 2, 3]
>>> l[0] = 0
>>> l
[0, 2, 3]
>>> d
{1: 'a', 2: 'b', 3: 'c'}
```

8.2.2 Dicts and Loops

Dicts are iterable. The *for* loop can be used to traverse a *dict* object, iterating through all the keys in the *dict*.

```
>>> d = {1: 'a', 2: 'b', 3: 'c'}
>>> for k in d:
...     print 'k={}, d[k]={}'.format(k, d[k])
...
k=1, d[k]=a
k=2, d[k]=b
k=3, d[k]=c
```

In the example above, the *dict d* is used as the container object in the *for* line, in the same way that *tuples* and *lists* are used in *for* lines. When a *dict* object is traversed, its keys are enumerated.

The method *iterkeys* can be used to achieve the same iteration:

```
>>> d = {1: 'a', 2: 'b', 3: 'c'}
>>> for k in d.iterkeys():
...     print k
...
1
2
3
```

The method *itervalues* can be used to iterate through the values that correspond to each key in a *dict*.

```
>>> d = {1: 'a', 2: 'b', 3: 'c'}
>>> for v in d.itervalues():
...     print v
...
a
b
c
```

As can be seen from the example above, if there are duplicate values in *d*, each is enumerated by the *for* loop.

As a third way of iterating through a *dict*, the method *iteritems* can be used to iterate through all the items in a *dict*, it is similar to the *enumerate* method for sequence objects, which enumerates both indices and values.

```
>>> d = {1: 'a', 2: 'b', 3: 'c'}
>>> for k, v in d.iteritems():
...     print 'd[{}]={}'.format(k, v)
...
d[1]=a
d[2]=b
d[3]=c
```

During the iteration over a *dict* object, the *dict* object cannot be modified. Such a modification can lead to a runtime error being raised.

```
>>> d = {1: 'a', 2: 'b', 3: 'c'}
>>>
>>> for k in d:
...     d[k+1] = 'd'
...
Traceback (most recent call last):
  File "<stdin>", line 1, in <module>
RuntimeError: dictionary changed size during iteration
```

Because *dict* objects are not ordered, the method *iterkeys*, *itervalues* and *iteritems* can iterate through the *dict* in arbitrary orders. In order to alow the specification of the order of traversal, a set of alternative methods, *keys*, *values* and *items* can be used, which return a copy of the keys, values and items of the *dict* object, respectively, as a *list* object. Because the *list* object is not a part of the *dict* object, it can be modified and reordered, before being used for traversal of the *dict*.

```
>>> d = {1: 'a', 2: 'b', 3: 'c'}
>>> l = d.keys()
>>> for k in l:
...     print k, d[k]
...
1 a
2 b
3 c
>>> l.reverse()
>>> for k in l:
...     print k, d[k]
...
3 c
2 b
1 a
```

8.2.3 Dicts and Functions

Similar to *lists*, *dicts* are mutable. As a result, *dict* arguments can be modified by a function. The function *filterdict* below takes a simple *dict* argument, and removes all the items with 0 values.

```
>>> def filterdict(d):
...     keys = d.keys()
...     for k in keys:
...         if d[k] == 0:
...           del d[k]
...
>>> d1 = {'a': 1, 'b': 2, 'c': 3, 'd': 4, 'e': 5, 'f':
    0}
>>> d2 = {True: 0, False: 1}
>>> d3 = {0: 1, 1: 2, 3: 2}
>>> filterdict(d1)
>>> d1
{'a': 1, 'c': 3, 'b': 2, 'e': 5, 'd': 4}
>>> filterdict(d2)
>>> d2
{False: 1}
>>> filterdict(d3)
>>> d3
{0: 1, 1: 2, 3: 2}
```

Given an input argument *d*, the function *filterdict* works by iterating through a copy of the keys in *d*, deleting those items with a value of 0. Because the local argument *d* is bound to the same object as *d*1, *d*2 and *d*3 in the calls *filterdict*(*d*1), *filterdict*(*d*2) and *filterdict*(*d*3), respectively, the global variables *d*1, *d*2 and *d*3 are modified by the function calls. Similar to *lists*, it is adviced not to modify a *dict* while iterating through it. The example makes a copy of the keys and iterates though the copy, rather than iterating though the *dict* directly.

The previous chapter introduced the use of a special *list* parameter, marked with *, to support an arbitrary number of optional argument. Python additionally allows the use of a **special dict parameter**, marked with **, to support an arbitrary number of keyword arguments in function calls. When a special argument with ** is given in the definition of a function, all keyword arguments that cannot be explicitly matched in a call to the function are packed into a *dict* and assigned to the special parameter, with the keys being the string form of the keywords, and the corresponding values being the values of the input arguments.

For example,

```
>>> def f(a, b, **d):
...     print 'explicit arguments'
...     print "a =", a
...     print "b =", b
...     print 'packed arguments'
...     for k in d:
...         print k, '=', d[k]
...
>>> f(1, 2, c=3, d=4, e=5)
```

```
explicit arguments
a = 1
b = 2
packed arguments
c = 3
e = 5
d = 4
>>> f(a=0, b=-1, e=-5)
explicit arguments
a = 0
b = -1
packed arguments
e = -5
```

In the example above, the function *f* takes two explicit arguments *a* and *b*, and a special *dict* argument *d*, printing out all the explicitly and implicitly matched arguments and their values. The outputs show that the arguments *a* and *b* are always matched explicitly, while any additional keyword arguments are packed into *d*.

The special *dict* argument and the special *list* argument can be used simultaneously, accepting keyword arguments and non-keyword arguments that cannot be explicitly matched, respectively.

```
>>> def f(a, b, *l, **d):
...     print 'Explicit arguments'
...     print "a =", a
...     print "b =", b
...     print 'List arguments'
...     for i in l:
...         print i
...     print 'Dict arguments'
...     for k in d:
...         print k, '=', d[k]
...
>>> f(1, 2)
Explicit arguments
a = 1
b = 2
List arguments
Dict arguments
>>> f(1, 2, 3)
Explicit arguments
a = 1
b = 2
List arguments
3
Dict arguments
>>> f(1, b=2, c=3)
Explicit arguments
a = 1
b = 2
List arguments
Dict arguments
c = 3
>>> f(1, 2, 3, d=4, c=5, f=6)
Explicit arguments
```

```
a = 1
b = 2
List arguments
3
Dict arguments
c = 5
d = 4
f = 6
```

Similar to the case of *lists*, a *dict* can be passed as keyword arguments to function by using the prefix **.

```
>>> def f(a,b):
        return a+b

>>> d={'a':1,'b':2}
>>> f(**d)
3
```

In the example above, a call to $f(**d)$ is effectively the same as a call to $f(a = 1, b = 2)$.

8.3 Sets and Bitwise Operations

Another commonly-used container type is *set*, which represents an unordered container of immutable objects. Similar to *lists* and *dicts*, *sets* are mutable. Compared with *lists*, *sets* have two main differences. First, *sets* are unordered but *lists* are ordered. Second, *sets* do not allow duplicated items, but the same item can exit in a *list* multiple times. *Sets* are connected to *dicts* in that both *set* items and *dict* keys are distinct. A *set* can be implemented as a special *dict*, with the keys representing the items in the *set*, and the values being dummy values (e.g. 1), which are ignored.

A **set literal** consists of a comma-separated list of literals or identifiers, enclosed in a pair of curly brackets.

```
>>> a = 1
>>> i = 5.6
>>> s = {a, True, i}
>>> s
set([5.6, True])
>>> type(s)
<type 'set'>
```

The literal of an empty *set* is *set*() rather than {}, which represents an empty *dict* object. The *set* function is similar to the functions *int*, *float*, *complex*, *bool*, *str*, *tuple*, *list* and *dict* in that it constructs a new *set* object.

Set operators. *Sets* support the *in* and *not in* operators, which return whether an item is in a *set* or not. The builtin function *len* can be used to find the cardinality of a *set*.

```
>>> s = {1, 2, 3}
>>> 1 in s
True
>>> 3 not in s
False
>>> 4 not in s
True
>>> len(s)
3
>>> len(set()) # empty set
0
```

The operators >, >=, < and <= are applied to two *set* objects, returning whether the first *set* is a proper superset, a superset, a proper subset and a subset of the second *set*, respectively. The methods *issubset* and *issuperset* achieve the same functionalities as the operators >= and <=, respectively.

```
>>> a = {1, 2, 3}
>>> b = {3, 4, 5}
>>> c = {3}
>>> a >= c
True
>>> a > b
False
>>> b.issuperset(c)
True
>>> set().issubset(c)
True
>>> c < b
True
```

The following operators and methods are applied to two *set* objects, returning a new *set* object.

- The | operator and the *union* method return the union of the two *sets*;
- The & operator and the *intersection* method return the intersection of the two *sets*;
- The − operator and the *difference* method return the difference between the two *sets*, which contains all the items in the first *set* that are not in the second *set*.
- The ∧ operator and the *symmetric_difference* method return the symmetric difference between the two *sets*, which consists of elements in either but not both of the *sets*.

```
>>> s1 = {1, 2, 3}
>>> s2 = {3, 4}
>>> s1&s2
set([3])
>>> s1.union(s2)
set([1, 2, 3, 4])
>>> s1-s2
set([1, 2])
>>> s1.symmetric_difference(s2)
set([1, 2, 4])
```

Conversion between sets and other types. *Set* objects can be converted into strings, Boolean objects, *tuples* and *lists*. The string conversion of a *set* object is a literal form of the *set*; the Boolean conversion of a *set* object is *False* only if the *set* is empty, and *True* otherwise; the *tuple* and *list* conversions of a *set* object consists of all the items in the *set*, in arbitrary (but not random) order.

```
>>> s = {'a', 'b', 'c', 1, 2, 3}
>>> str(s)
"set(['a', 1, 2, 3, 'c', 'b'])"
>>> print s
set(['a', 1, 2, 3, 'c', 'b'])
>>> bool(s)
True
>>> if s:
...     print 'non-empty'
...
non-empty
>>> tuple(s)
('a', 1, 2, 3, 'c', 'b')
>>> list(s)
['a', 1, 2, 3, 'c', 'b']
```

Strings, *tuples*, *lists* and *dicts* can be converted into *sets*. The *set* conversion of a string consists of all the characters in the string, the *set* conversion of a *tuple* or a *list* consists of all the items in the sequence, while the set conversion of a *dict* consists of all the keys.

```
>>> set("123")
set(['1', '3', '2'])
>>> set(['c', True, 1+3j])
set([True, 'c', (1+3j)])
>>> set({1:1, 2:0, 3:-1})
set([1, 2, 3])
```

8.3.1 Set Modification

The *set* type supports a set of methods that modify a *set* object, which can be classified into *set*-element methods and *set-set* methods. Commonly used *set*-element methods include:

- *add(x)*, which takes an immutable argument *x* and adds the argument into the *set*;
- *remove(x)*, which takes an immutable argument *x* and removes the argument from the *set*. A *key error* is raised if the argument is not in the *set*;
- *discard(x)*, which takes an immutable argument *x* and removes the argument from the *set*. If the argument is not in the *set*, no action is taken.
- *pop()*, which takes no argument, and removes an arbitrary item from the *set*, returning its value.

```
>>> s = {1, 2, 3}
>>> s.add(0)
>>> s
set([0, 1, 2, 3])
>>> s.remove(5)
KeyError: 5
>>> s.discard(3)
>>> s
set([0, 1, 2])
>>> s.pop()
0
>>> s
set([1, 2])
```

Commonly-used *set-set* methods include:

- *update(s)*, which takes a *set* argument *s*, adding all the items in the argument *s* into the *set*. The operator |= achieves the same functionality as this method.
- *intersection_update(s)*, which takes a *set* argument *s*, and updates the *set*, keeping only those items that are also in the argument *s*. The operator &= achieves the same functionality as this method.
- *difference_update(s)*, which takes a *set* argument *s*, and updates the *set*, removing those items that are also in the argument *s*. The operator −= achieves the same functionality as this method.
- *symmetric_difference_update(s)*, which takes a *set* argument *s*, and updates the *set*, keeping only items that are in either the *set* or the argument *s*, but not in both. The operator ∧= achieves the same functionality as this method.

```
>>> s1 = {1, 2, 3}
>>> s2 = {3, 4}
>>> import copy
>>> s = copy.copy(s1)
>>> s.update(s2)
>>> s
set([1, 2, 3, 4])
>>> s = copy.copy(s1)
>>> s &= s2
>>> s
set([3])
>>> s = copy.copy(s1)
>>> s |= s2
>>> s
set([1, 2, 3, 4])
>>> s = copy.copy(s1)
>>> s.difference_update(s2)
>>> s
set([1, 2])
>>> s = copy.copy(s1)
>>> s ^= s2
>>> s
set([1, 2, 4])
```

In the example above, the four *set-set* operations are applied to the *sets* {1, 2, 3} and {3, 4}, respectively. Because they are set modification methods, a copy of the original *set* is necessary for keeping the original *sets*. The *set* type provides a method, *copy*, which takes no argument and returns a copy of the *set*. As a result, $copy.copy(s1)$ in the examples above can be replaced with $s1.copy()$.

Similar to *dicts*, *sets* support the *clear* method, which takes no arguments and removes all the items from the *set*.

Set are iterable. The *for* loop can be used to iterate through a *set* object. Because *sets* are unordered, their iteration takes an arbitrary (but not random) order.

```
>>> s = {1, 'a', 2, 'b', None, True}
>>> for i in s:
...     print i
...
a
True
2
b
None
```

8.3.2 Bitsets and Bitwise Operators

Sets are inherently connected to binary numbers. As mentioned in Chap. 2, integers can be written in binary forms, in which each digit is either 0 or 1. Intuitively, the values 0 and 1 can be used to represent whether an item is in a set or not. As a result, if the possible elements in a set is finite, they can be indexed and associated with a finite number of bits in a binary number. Each possible set element corresponds with a specific bit in the number. The correlation between a finite set and a binary number is shown in Fig. 8.1.

As can be seen from the figure, the value (i.e. 0 or 1) of the ith bit in the integer corresponds to the presence of the ith possible item in the set. On the other hand, the ith bit in the integer also represents the value of 2^i. As a result, the set $\{e1, e3, e7\}$ corresponds to the integer $2^1 + 2^3 + 2^7 = 2 + 8 + 128 = 138$.

Python supports binary integer literals, which start from the letters '0b', followed by binary digits. For example, *0b1* represents $2^0 = 1$, *0b10* represents $2^1 = 2$, and *0b100* represents $2^2 = 4$. *0b101* represents $2^2 + 2^0 = 5$.

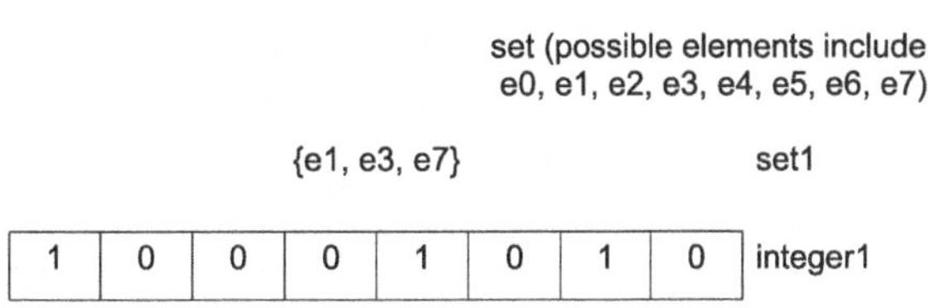

Fig. 8.1 The correlation between finite sets and binary numbers

```
>>> 0b1101  # 2**3+2**2+2**0=13
13
>>> 0b1011  # 2**3+2**1+2**0=11
11
```

The bulitin function *bin* takes an integer argument and returns a string that shows the binary form of the number.

```
>>> bin(13)
'0b1101'
>>> bin(125)
'0b1111101'
```

The connection between binary numbers and sets allows the use of integers to implement or represent sets, provided that possible items in the set can be indexed. For example, the set $\{e1, e3, e7\}$ in Fig. 8.1 can be represented by the integer *0b10001010* $= 2^7+2^3+2^1= 138$. Given an integer, its corresponding set can be examined by checking item presence in its binary representation.

The union, intersection and symmetric difference operations between two sets are connected to digit-wise (or bitwise) Boolean operations between two binary numbers. For example, in the union of the sets *s1* and *s2*, an item is present if the item is present in *s1 or* in *s2*, while in the intersection of *s1* and *s2*, an item is present if the item is present in *s1 and* in *s2*. As a result, the binary numbers that correspond to the union and intersection of *s1* and *s2* can be calculated from the binary numbers that correspond to *s1* and *s2* by the *bitwise or* and *bitwise and* operations, respectively. The binary number that corresponds to the symmetric difference between the two sets *s1* and *s2* can be calculated from the binary numbers that correspond to *s1* and *s2* by the *bitwise xor* operation. The three bitwise operations are defined as follows.

In a *bitwise or* operation between two numbers *n1* and *n2*, the *i*th bit of the result is 1 if the *i*th bit of *n1* is 1 *or* the *i*th bit of *n2* is 1; in a *bitwise and* operation between two numbers *n1* and *n2*, the *i*th bit of the result is 1 if the *i*th bit of *n1* is 1 *and* the *i*th bit of *n2* is 1; in a *bitwise xor* operation between two numbers *n1* and *n2*, the *i*th bit of the result is 1 if the *i*th bit of *n1* is different from the *i*th bit of *n2*, and 0 otherwise. Python supports the bitwise operations above by the integer operation |, & and $\wedge$, respectively.

```
>>> a = 0b1101
>>> b = 0b1001
>>> a
13
>>> b
9
>>> a|b
13
>>> bin(a|b)
'0b1101'
>>> a&b
```

```
9
>>> bin(a&b)
'0b1001'
>>> a^b
4
>>> bin(a^b)
'0b100'
```

The correlation between set operations and bitwise integer operations is shown in Fig. 8.2. Note that the *difference* (i.e. –) operation between two sets is not correlated to the – operation between two numbers, which is an arithmetic operator rather than a bitwise operator.

|, & and ∧ are all binary bitwise operations, taking two operands. There is a unary bitwise Boolean operator, ~ , which takes one operand and flips every bit in it:

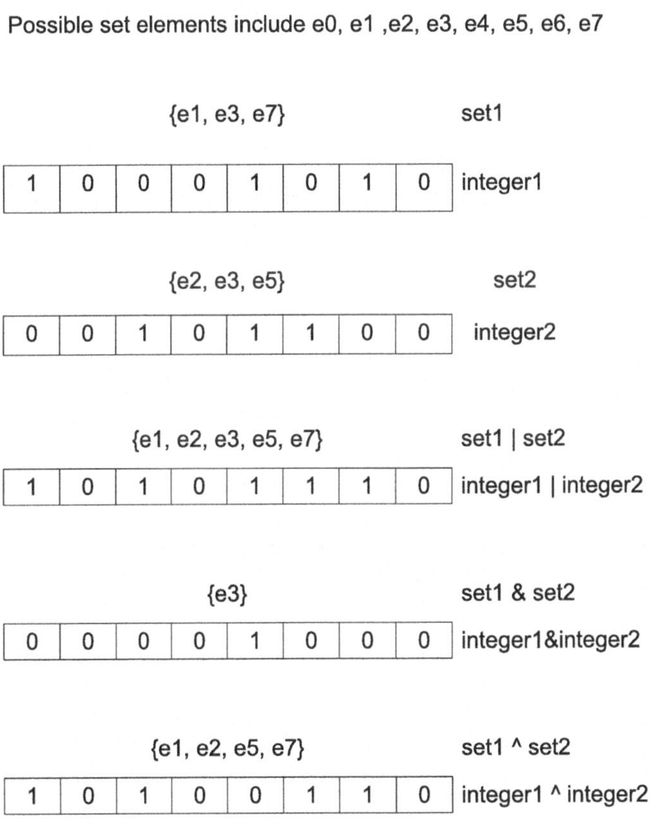

Fig. 8.2 The correlation between set operations and bitwise integer operations

```
>>> a=0b1010101100
>>> bin(~a)
'-0b1010101101'
```

In addition to bitwise Boolean operations, Python also supports a set of bitwise shifting operations between two integers. In particular, *right-shift* operation $n \gg k$ shifts the binary form of n to the right by k bits, and the *left-shift* operation $n \ll k$ shifts the binary form of n to the left by k bits. In terms of the value, the operation $n \gg k$ is equivalent to $n/(2^k)$ and the $n \ll k$ operation is equivalent to $n \times 2^k$. In both $n/(2^k)$ and $n \times 2^k$, n is treated as a positive integer.

```
>>> n = 0b1011
>>> bin(n<<1)
0b10110
>>> bin(n>>1)
0b101
>>> bin(n>>2)
0b10
```

Exercises

1. Write a function

   (a) *is_valid(s)*, which takes a string argument *s*, representing a password, and return a Boolean value indicating whether the input password is valid. A valid password must contain at least 8 characters, with one character being numerical and one character being punctuation. (i.e. ' , ', ' . ', ' ! ', ' > ', ' < ' or ' ♯ ').
   (b) *dec(s)*, which is the reverse of *bin(x)*. It takes a string starting with 'ob', followed by a binary number, and returns the decimal value of the number.
   (c) Write a function, *lcp(s1,s2)*, which takes two string arguments $s1$, $s2$ and returns the largest common prefix of $s1$ and $s2$.
   (d) Write a function, *reverse_dic(d)*, which takes a *dict* argument *d*, and returns a new *dict d1*, whose keys are the unique values of *d*. The value of each key *k* in *d1* is a list that contains all the keys in *d* that has the value *k*, sorted in ascending order. For example:

   ```
   d={'d':1,'b':3,'c':1,'a':2,'e':1,'f':2}
   d1=reverse_dic(d)
   d1
   {1:['c','d','e'],2:['a','f'],3:['b']}
   ```

   (e) Write a function, *unique(t)*, which takes a *tuple* argument *t*, which contains integers, and returns a new *tuple* that contains all the unique items in *t*, sorted in descending order. (Hint: consider *sets*).
   (f) Write simple string compression functions. *compress(s)* takes a string argument and returns its compression, *decompress(s)* takes a compressed string argument and return the original string. The compression algorithm is simple, consecutive runs of the same character *c* for *n* times (*n*>2) are replaced with the sub string *cn*. For example, "aa bbb cccc"is compressed into 'aab3c4'

2. What are the values of the expressions below?

   (a) 3 & 5
   (b) 3 | 5
   (c) 3 ∧ 5
   (d) ~5
   (e) 3 and 5
   (f) 3 or 5
   (g) 3 ≫ 1
   (h) 3 ≪ 1

3. DNA sequences consist of four base types of molecules, which can be represented by the letters A, C, G and T, respectively. Write a function, *countbase*, which takes a string argument representing a DNA sequence, and returns a *dict* that contains the count of each base, with the base letter being the key and the count as the value. For example, *countbase*('ACGTGGGAGTC') returns *{'A':2, 'C':2, 'G':5, 'T':2}*, and *countbase*('ATBLC') returns *{'A':1, 'C':1, 'G':0, 'T':1}*.
4. Write a function to encode shapes into integers. In particular, there are four types of shapes to encode: triangles, rectangles, regular pentagons and regular hexagons. Each shape is drawn in a 128 × 128 grid, and has a color from red, green, blue, yellow and white.
   Use 2 bits to encode the shape, 0 = triangle, 1 = rectangle, 2 = regular pentagon and 3 = regular hexagon. Use 3 bits to encode the color, 0 = red, 1 = green, 2 = blue, 3 = yellow, 4 = white and 5/6/7 = invalid.
   Use a total of 42 bits to represent the coordinates. For triangles, the coordinates specify the three vertices, each taking 2 × 7 bits (0–127). For rectangles, 28 bits are used to specify the top-left vertex and the bottom-right vertex, each taking 2 × 7 bits. The last 2 × 7 bits are all zeros. For regular pentagons, the coordinates specify the top vertex and its two neighbors, each taking 2 × 7 bits. For regular hexagons, the coordinates specify the left three vertices, each taking 2 × 7 bits.
   As a result, each shape can be encoded into 26 bits. Write a function, *encodeshape*, which takes the shape, color and coordinates as input arguments, and returns the integer code. For example, *encodeshape*('triangle', 'yellow', 10, 10, 90, 10, 60, 58) returns 13540610186810, which is *00 011 0001010 0001010 1011010 0001010 0111100 0111010*. The first two bits 00 represent triangle, the next three bits 011 represent yellow, the next 2 × 7 bits, 0001010 and 0001010, represent the first vertex, (10, 10), and so forth.
   Write another function, *decodeshape*, which takes a number argument and returns the shape it encodes. For example, *decodeshape*(13540610186810) returns ('triangle', 'yellow', 10, 10, 90, 10, 60, 58).
5. *collections* is a useful module that provides extended collection types. There is one useful class called *defaultdict*, which assigns default values to missing keys automatically when they are accessed. For example,

```
>>> import collections
>>> d = collections.defaultdict(lambda:0)
>>> d['A'] = 1
>>> d['A']
1
>>> d['B']
0
>>> d['C'] += 1
>>> d['C']
1
```

The function call *collections.defaultdict* takes a function argument, and returns a *defaultdict* object. The default value of missing keys is the return value of the argument function.
In the example above, one *lambda* function that takes no arguments and returns 0 is used to construct a *defaultdict*.
Use *defaultdict* as the main data structure, and write a program that records the cities that certain people visit. The program repeatedly asks the user to enter sentences in the form of 'X visited Y.', where *X* is the name of a person and *Y* is a city. When the user enters an invalid sentence, a warning is given. When the user enters an empty sentence, the program exits and prints the cities that each person has visited. For example,

```
c:\user\yue_zhang> python visit.py
Enter sentence: John visited Paris.
Enter sentence: Mary visited Beijing.
Enter sentence: John visited Paris.
Enter sentence: Mary visited Tokyo.
Enter sentence: Hello!
Invalid sentence --- Enter 'X visited Y.'
Enter sentence: Mary visited London.
Enter sentence: John visited New York.
Enter sentence: Mary visited Beijing.
Enter sentence:

John has visited New York and Paris.
Mary has visited Beijing, London and Tokyo.
c:\user\yue_zhang>
```

6. Write a function, *split(s)*, which takes a string argument *s*, representing a piece of text, and returns a *dict* object, representing the sentences in the text. The keys to the dict are sentence indices, starting from 1. For example.

```
s="I bought a new laptop for $3.5k. The medel
    was recommended by my friend, Dr. Smith. I
    like its functionalities, but found that the
     same medel is sold for only $2.8k at amazon
    .com... Lessons learned? It's worth checking
     online prices before shopping!"
d=split(s)
d
{1:"I bought a new laptop for $3.5k.",
 2:"The medel was recommended by my friend,
    Dr. Smith."
```

```
3:"I like its functionalities, but found that
    the same medel is sold for only $2.8k at
    amazon.com..."
4:"Lessons learned?"
5:"It's worth checking online prices before
    shopping!"}
```

Hint: There are three boundary punctuations, '.', '?', '!'. However, ambiguities exist in many cases. For example, '.' is typically a sentence boundary marker when followed by a space, but special terms such as 'Dr' also ends with '.'. As a result, no program can perfectly handle all cases. Try to make the function work in most situations.

Chapter 9
Problem Solving Using Lists and Functions

This chapter discusses problem solving using nested loops and functions. Typical problems with *lists* of *lists* are discussed, and common matrix operations are illustrated. Recursive functions and functional programming concepts are introduced, by showing how standard problems introduced in Chap. 5 can be solved using the new methods. The last section of this chapter introduces I/O with files and URL web pages, showing how strings and *dicts* can be used to organize text data, and how the *pickle* module can be used to serialized arbitrary Python objects into files.

9.1 Lists of Lists and Nested Loops

Many types of real-world data can be represented by a *list* of *lists*. For example, the final examination results of a class can be represented by a *list* of *lists*, where each sub *list* consists of the scores of 4 different subjects of a particular student. The trading records of a stock market can also be represented by a *list* of *lists*, where each sub *list* consists of the daily closing prices of a particular stock. In mathematics, matrices can also be represented by *lists* of *lists*, where each sub *list* represents a particular row.

There are two general ways to solve problems that involve a *list* of *lists*. The first is to treat each sub *list* as an atomic unit, and thereby process the *list* of *lists* in the same way as processing a *list* of other objects. The second way is to use a nested loop to iterate over the items in the *list* of *lists*. Using the financial examination results as an example, this section discusses the two types of solutions in details.

9.1.1 Treating Sublists as Atomic Units

Consider the problem of finding the total scores of the four subjects for each student in the final examination. The task is to write a function that that takes a *list* of *lists* as the only argument, and returns a list that contains the sum of each sub *list*.

Y. Zhang, *An Introduction to Python and Computer Programming*,
Lecture Notes in Electrical Engineering 353, DOI 10.1007/978-981-287-609-6_9

The problem can be solved by treating each sub *list* as an atomic unit, using the built-in function *sum* to calculate its total value. As a result, the idea of generalized summation can be applied to solve the problem: the return value is initialized to an empty *list*, and incrementally update by iterating through the input *list*. When each sub *list* is enumerated, its total value is added to the back of the return value. After the input *list* is exhausted, the return value contains the sum of each sub *list*. This solution can be written as the *sumlists* function below.

```
>>> def sumlists(l):
...     s = []
...     for subl in l:
...         s.append(sum(subl))
...     return s
...
>>> sumlists([ [90, 95, 100, 100], [80, 70, 65, 71],
    [95, 60, 75, 55] ])
[385, 286, 285]
```

Now suppose that the problem is to find the average score of each student instead of the total score. The function *sumlists* can be modified slightly, calculating the average of each sub *list* by dividing the sum the number of items in the sub *list*. The function *avglists* below is an example implementation.

```
>>> def avglists(l):
...     s = []
...     for subl in l:
...         s.append(float(sum(subl))/len(subl)) #
    average
...     return s
...
>>> avglists([ [90, 95, 100, 100], [80, 70, 65, 71],
    [95, 60, 75, 55] ])
[96.25, 71.5, 71.25]
```

The expression *float(sum(subl))/len(subl)* may be difficult to understand. Alternatively, it can be modularized and implemented as a function. The example below uses the function *avg* to calculate the average of a *list* of numbers. It can make the *avglists* function easier to understand compared to the implementation of *avglists* above.

```
>>> def avg(l):
...     s = 0
...     for i in l:
...         s += i
...     return float(s) / len(s)
...
>>> def avglists2(l):
...     s = []
...     for subl in l:
...         s.append(avg(subl))
...     return s
...
>>> avglists2([ [90, 95, 100, 100], [80, 70, 65, 71],
    [95, 60, 75, 55] ])
[96.25, 71.5, 71.25]
```

Now suppose that one needs to calculate the averaged score of all the subjects of all the students. The problem is to find the average of all the items of all the sublists in a *list* of *lists*. It can be solved by calculating the sum of all the items of all the sub *lists*, and then dividing the sum by the total number of items of all the sub *lists*. The function *avglistoflists* below is an example implementation.

```
>>> def avglistoflists(l):
...     s = 0
...     c = 0
...     for subl in l:
...         s += sum(subl)
...         c += len(subl)
...     return float(s) / c
...
>>> avglistoflists([ [90, 95, 100, 100], [80, 70, 65,
    71], [95, 60, 75, 55] ])
79.6666666667
```

The function *avglistsoflists* above uses two variables, *s* and *c*, to maintain the sum and count of the items of the sub *lists*, respectively. The input *list* is iterated over so that *s* and *c* are updated incrementally. In this process, each sub *list* is treated as an atomic unit, processed by a function.

As mentioned in the beginning of the section, an alternative solution to *list* of *lists* problems is to iterate through all the sub *list* items using nested loops. A nested loop can consist of two layers. The **outer loop** iterates through the *list*, enumerating each sub *list*. The inner loop iterates through a sub *list*, enumerating its items. In one execution of a nested loop, the outer loop is executed once, while the inner loop is executed once at each iteration of the outer loop.

For example, consider rewriting the *avglists2* function above, replacing the function call *avg* in the function body with the function body of *avg*.

```
>>> def avglists3(l):
...     s = []
...     for subl in l:
...         subs = 0                               # avg
...         for i in subl:                         # avg
...             subs += i                          # avg
...         suba = float(subs) / len(subl) # avg
...         s.append(suba)                         # avg
...     return s
...
>>> avglists3([ [90, 95, 100, 100], [80, 70, 65, 71],
    [95, 60, 75, 55] ])
[96.25, 71.5, 71.25]
```

The functional *avglists3* is different from the function *avglists2* only in the loop body. *avglists3* uses an inner loop to calculate the average of each sub *list*. Given a sub *list subl*, it uses the variable *subs* to keep the sum of *subl*, and the variable *suba* to keep its average. *subs* is initialized to 0, and updated incrementally by iterating through *subl*. *suba* is calculated by dividing *subs* by the length of *subl*. The five lines of statements commented with *avg* calculate *subs* and *suba* from a given *subl*, regardless of the rest of the control flow. The five lines of code are put inside the *for* loop over *l*, and repeatedly executed with *subl* being each sub *list* in *l*. As a result, the *for* loop over *subl* plays the role of an inner loop.

Nested loops can be used to iterate through all the items in all the sub *lists* in a *list* of *lists*. In therefore provides an intuitive solution to the problem of finding the averaged score of the class. The function *avglistoflists2* is an alternative implementation of *avglistoflists* using nested loops.

```
>>> def avglistoflists2(l):
...     s = 0
...     c = 0
...     for subl in l:
...         for i in subl:   # i enumerates each sub list item
...             s += i
...             c += 1
...     return float(s) / c
...
>>> avglistoflists2([ [90, 95, 100, 100], [80, 70, 65, 71],
    [95, 60, 75, 55] ])
79.6666666667
```

In the example above, *s* and *c* are used to hold the sum and count of items of all sub *lists*, respectively. A call to *avglistoflist2* executes the outer loop (i.e. *for subl in l*) once. The body of the outer loop is repeated three times, with *subl* being [90, 95, 100, 100], [80, 70, 65, 71] and [95, 60, 75, 55], respectively. Each time the outer loop body is executed, the inner loop (*for i in subl*) is executed once, with the loop body being repeated four times. As a result, the statement *c += 1* in the inner loop body is executed 12 times in total. Each time *i* being assigned the value of a different sub *list* item.

Nested loops for item combination. In addition to enumerating through sub *list* items in a nested *list*, a second typical use of nested loops is to find all possible combination of items in different *lists*. For example, the function *combine* below takes two *list* arguments, and return a *list* that contains all possible combinations of their items.

```
>>> l1=[1,2,3,4,5]
>>> l2=['a','b','c']
>>> def combine(x,y):
...       retval=[]
...       for i in x:
...           for j in y:
...                 retval.append((i,j))
...     return retval
...
>>> combine(l1,l2)
[(1, 'a'), (1, 'b'), (1, 'c'), (2, 'a'), (2, 'b'),
(2, 'c'), (3, 'a'), (3, 'b'), (3, 'c'),
(4, 'a'), (4, 'b'), (4, 'c'), (5, 'a'), (5, 'b'),
(5, 'c')]
```

In the example above, the outer loop enumerates the items in the first *list*, when each item is enumerated, the inner loop is executed once, enumerating the items in the second *list*. As a result, each item combination is enumerated once and only once, by the nested loop.

Loops can be nested in more than two levels, for the solution of more complex problems. The analysis and design of three-layer nested loops is similar to those of two-layer nested loops, and it is important to understand the functionality of the inner loop, so that it can be used correctly in combination of the outer loop. In case the program structure is too complex to understand, nested loops can be simplified by the use of function calls to replace inner loops, as illustrated by the contrast between *avglists2* and *avglists3*.

9.1.2 Matrices as Lists of Lists

Matrices can be represented by *lists* of *lists*, with each sub *list* representing a row in the matrix. For example, the matrix $M = \begin{vmatrix} 1 & 2 & 3 & 4 \\ 5 & 6 & 7 & 8 \\ 9 & 10 & 11 & 12 \end{vmatrix}$ can be represented by the *list*

```
>>> m = [ [1, 2, 3, 4],
...       [5, 6, 7, 8],
...       [9, 10, 11, 12] ]
```

Using this formulation, the *i*th row of *m* can be accessed by the expression *m[i-1]*:

```
(continued from above)
>>> m[0] # the first first row
[1, 2, 3, 4]
>>> m[2] # the third row
[9, 10, 11, 12]
```

The item $m_{i,j}$ can be accessed by *m[i-1][j-1]*:

```
(continued from above)
>>> m[0][0] # the first row, the first column
1
>>> m[2][1] # the third row, the second column
10
```

As nested *lists*, matrices can be processed using loops. The following examples give a set of functions for matrix manipulation.

The functions *get_row* and *get_column* take a matrix argument, and additionally an integer argument that specifies an index i, and return a copy of the ith row and column of the matrix as a *list*, respectively.

```
(continued from above)
>>> def get_row(m, i):
...     return m[i][:]
>>> def get_column(m,i):
...     s = []
...     for row in m:
...         s.append(row[i])
...     return s
...
>>> get_row(m,1)
```

```
[5, 6, 7, 8]
>>> get_column(m,2)
[3, 7, 11]
```

The function *get_column* above works by initializing the return value to an empty *list*, and updating it by iterating through each row in the matrix. At each iteration, the item in the specified column is added to the back of the result. After all the rows have been enumerated, the return value contains the specified column.

A simple matrix operation is *scaling*. Given a matrix M and a scalar a, aM results in a matrix N, which has the same width and height as M, and satisfies $N_{i,j} = aM_{i,j}$ for all i, j. Matrix scaling can be performed by initializing the result as a copy of M, and then scaling each element by traversal of the new matrix.

```
(continued from above)
>>> def scale(a,M):
...     import copy
...     N=copy.deepcopy(M)
...     for i in range(len(N)): # row
...         for j in range (len(N[i])): # col
...             N[i][j]= N[i][j]*a # setitem
...     return N
...
>>> M = [[1,2,3], [4,5,6],[7,8,9]]
>>> scale(2,M)
[[2, 4, 6], [8, 10, 12], [14, 16, 18]]
>>> scale(-0.5,M)
[[-0.5, -1.0, -1.5], [-2.0, -2.5, -3.0], [-3.5, -4.0,
   -4.5]]
>>> M # value not changed
[[1,2,3], [4,5,6],[7,8,9]]
```

There are two things to note on the function *scale* above. First, *copy.deepcopy* is used to make a copy of M. Because M is a *list* of *lists*, it needs copied recursively to avoid sharing of sub *lists* between the copies. Second, iteration over N is based on indices rather than values, because modification of sub *lists* (i.e. rows) requires *setitem*, which uses indices.

Another common matrix operation is *addition*, which adds two matrixes of the same size by each element. Matrix addition can be performed by initializing the result as one input matrix, and then updating each element by addition with the corresponding element in the other input matrix.

```
(continued from above)
>>> def add(M,N):
...     import copy
...     R = copy.deepcopy(M)   # initialize
...     for i in range(len(R)):    # row
...         for j in range(len(R[i])):   # col
...             R[i][j]+= N[i][j]
...     return R
...
>>> M=[[1,0,0], [0,1,0], [0,0,1]]
>>> N=[[1,2,3], [4,5,6], [0,0,0]]
>>> add(M,N)
```

```
[[2, 2, 3], [4, 6, 6], [0, 0, 1]]
>>> M
[[1, 0, 0], [0, 1, 0], [0, 0, 1]]
>>> N
[[1, 2, 3], [4, 5, 6], [0, 0, 0]]
```

Traversal by index can also be used to perform the *transposition* operation. Given a matrix M with width w and height h, the transpose M^{T} has width h and height w, and satisfies $M^{\mathrm{T}}_{i,j} = M_{j,i}$ for all i, j.

```
(continued from above)
>>> def transpose(M):
        #initialize w*h matrix
...     N=[]
...     for i in range(len(M[0])):
...         N.append([0]*len(M))
        # fill items
...     for i in range(len(M[0])):
...         for j in range(len(M)):
...             N[i][j]= M[j][i]
        #return new value
...     return N
...
>>> M=[[1,2], [3,4], [5,6]]
>>> transpose(M)
[[1, 3, 5], [2, 4, 6]]
>>> M
[[1, 2], [3, 4], [5, 6]]
```

The function *transpose* initializes the output N using a loop. N is initialized as an empty *list*, and then updated by appending h zero-valued rows, each consisting of w zeroes. Here w and h are the width and height of the input matrix, respectively.

Note that N cannot be initialized by the expression $[[0] * h] * w$, by the reason discussed in Chap. 7 — the object $[0] * h$ is a *list*, which is mutable. Making w copies of $[0] * h$ will result in all the rows in N sharing the same *list* (i.e. $[0] * h$), rather than having w copies of the *list*.

The function *transpose* works by generalized summation. The result is initialized as all zeros, and then iterating updated by iterating through each index i, j, assigning *M[j][i]* to *N[i][j]*. Note that the nested loop can be viewed as an enumeration of all possible item combinations of $[0, 1, \cdots, len(M[0]) - 1]$ and $[0, 1, \cdots, len(M) - 1]$.

Matrix multiplication is one of the most difficult operations. Multiplying two matrixes M and N results in a matrix R, in which the item $R_{i,j}$ is the dot product of the ith row of M and the jth column of N. Matrix multiplication can be written by slightly modifying the previous example of matrix transposition, using the dot product operation $M_i \cdot N_j$ instead of $M_{i,j}$ as the incremental operation.

```
(continued from above)
>>> def dot_product(v1,v2):
...     s=0
...     for i in range(len(v1)):
...         s+=v1[i]*v2[i]
...     return s
```

```
...
>>> def multiply (M,N):
...     # initialize result matrix
...     R=[]
...     for i in range(len(M)):
...         R.append([0]*len(N[0]))
...     # change each value
...     for i in range (len(M)):
...         for j in range (len(N[0])):
...             R[i][j]=dot_product(get_row(M,i),
                                   get_column(N,j))
...     # return result metric
...     return R
...
>>> M=[[1,2],[3,4],[5,6]]
>>> N=[[1,0,1,-1],[0,1,1,1]]
>>> multiply (M,N)
[[1, 2, 3, 1], [3, 4, 7, 1], [5, 6, 11, 1]]
>>> M
[[1,2],[3,4],[5,6]]
>>> N
[[1,0,1,-1],[0,1,1,1]]
```

9.2 Functions and Problem Solving

Several problem solving techniques make use of function calls. For example, a type of problems can be solved by *recursion*, which works by recursively reducing it to smaller problems of the same nature. *Recursion* is typically implemented by *recursive function calls*.

Functions are also the key element in *functional programming*, a programming paradigm that is *declarative*, or description-based, rather than *imperative*, or action-based. Compared to imperative programs, functional programs can be easier to understand because of their declarative nature. For example, no complex dynamic control flow are involved in functional programs.

Python supports functional programming vis a set of special functions and statements. Although functional programming is not central to Python, and there has been debate on whether its support should be a core part of the language, knowledge of functional programming can often help in writing simpler and better programs.

Both recursive function calls and functional programming take root in the research of computing machines. In the early days of computer science, many theoretical computing machines were proposed, including the *Turing machine*, *recursive functions*, *λ-calculus* and *production systems*. It has been commonly believed that all these computing machines are equally powerful. Modern computers largely evolve from Turing machines, which are programmable (i.e. it is possible to separate programs and computers, so that different functionalities can be achieved by running different programs on the same computer), while the other computing machines above influence different programming paradigms for modern computers.

9.2.1 Recursive Function Calls

Recursive function calls are calls to a function by itself. For example, the following function contains a recursive call to itself, which leads to a memory error if not interrupted.

```
>>> def f(x):
...     print x
...     f(x+1)
...
>>> f(1)
1
2
3
4
5
6
7
^C
KeyboardInterrupt
```

A call to the function f above leads to a dead-loop execution sequence, which terminated by keyboard interruption. The execution sequence is shown in Fig. 9.1. Initially, IDLE calls $f(1)$, which leads to the execution of the function body. After printing the value of the argument x, which is 1, the function call $f(2)$ is executed. The same process repeats, with increasing levels of recursive calls, each having a local binding table. On the other hand, the function call $f(1)$ will not finish until the evaluation of the call $f(2)$ finishes, and the call $f(2)$ will not finish until the call $f(3)$ finishes. As a result, the recursion continues infinitely, with increasing memory consumption.

In order to avoid such infinite execution, recursive calls must be conditional. Under certain input, the function call should return directly, without making a further recursive call. For example, the function f above can be modified so that it returns directly if x is greater than 3.

```
>>> def f(x):
...     if x>3:
...         return
...     print x
```

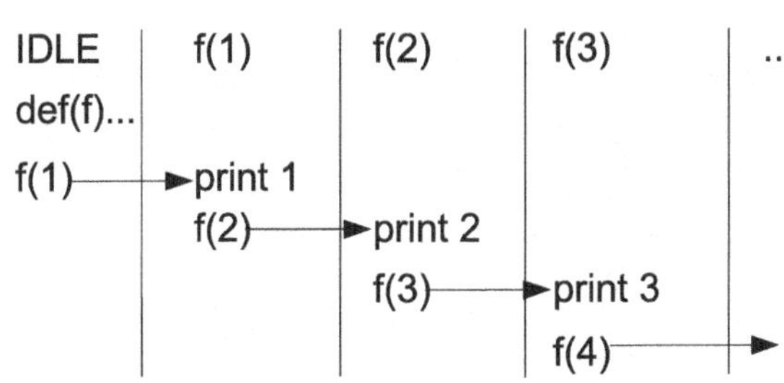

Fig. 9.1 Illustration of recursive function call

```
...        f(x+1)
...
>>> f(1)
1
2
3
```

The function call *f(1)* above finishes within finite steps, because no further recursive calls are made when $x = 4$. After *f(4)* returns, the execution of *f(3)* resumes with the expression *f(x+1)* being evaluated to the return value of the *f(4)* call, which is *None*. Because there are no more statements in the function body, *f(3)* return *None* by default. Similarly, *f(2)* and *f(1)* return in sequence. The condition under which no further recursion is made is also called the *stopping condition*. A recursive function definition must ensure that all possible recursive call sequences can reach the stopping condition. Otherwise, infinite recursion can happen, leading to overflow of memory.

Recursive function calls can be used to solve problems that can be reduced to smaller problems of the same nature. The problem of computing number sequences from difference equations introduced in Chap. 5 belongs to such problems. Given a *base case*, which can be solved directly, and a *difference equation*, which defines the correlation between two consecutive numbers, recursive function definitions offer a solution that is intuitive to understand. In such a solution, the base case serves as the stopping condition, and the difference equation is mapped into recursive calls.

For example, consider again the problem of finding the factorial of an integer. The base case is $0! = 1$, and the difference equation is $n! = (n-1)!n$. A recursive function definition can be used for this task.

```
>>> def factorial(n):
...     if n==0:  # basis
...         return 1
...     return factorial(n-1)*n
...
>>> factorial(5)
120
```

The dynamic execution sequence of the function call *factorial(5)* is shown in Fig. 9.2, where a sequence of nested recursive calls are executed, until the stopping condition is satisfied. The stack of recursive calls finishes in the last-in-first-out order, each callee passing its return value back to its caller. Note that Fig. 9.2 is different from Fig. 9.1 in setting a stopping condition, so that the recursion stops at *factorial(0)* instead of continuing endlessly.

For a second example, consider again the problem of finding the *n*th Fibonacci number. The base cases are $f_0 = 1$ and $f_1 = 1$, and the difference equation is $f_k = f_{k-1} + f_{k-2}$. The calculation of f_n can be written as a recursive function:

```
>>> def fib(n):
...     if n==0 or n==1:  # base
...         return 1
...     return fib(n-1)+fib(n-2)
...
```

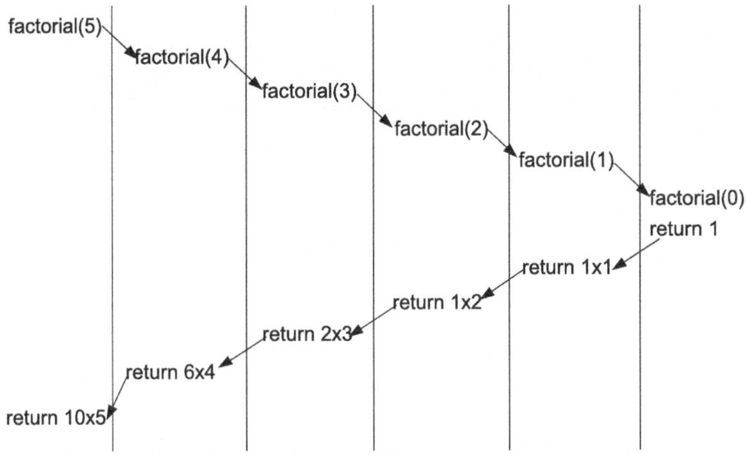

Fig. 9.2 Illustration of recursive function for calculating factorial

```
>>> fib(3)
3
>>> fib(5)
8
```

Executions of the *fib* function above give the same results as the *fib* program in Chap. 5. The sequence of recursive function calls in the evaluation of *fib(5)* is shown in Fig. 9.3. A call to *fib(5)* requires the values of *fib(4)* and *fib(3)* to be computed, respectively (step1 in the figure). The evaluation of *fib(4)* further leads to the evaluation of *fib(3)* and *fib(2)* (step2). The evaluation of *fib(3)* further leads to the evaluation of *fib(2)*and *fib(1)* (step3). The recursive process continues until the evaluation of *fib(1)* and *fib(0)*, which return directly (steps, 5, 6 and 7). After steps 5 and 6 are completed, step 4 is completed with *fib(2)* being returned. Similarly, after steps

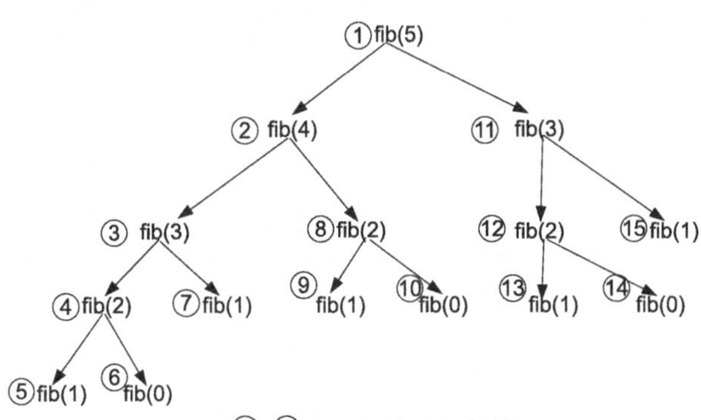

Fig. 9.3 Illustration of recursive function for Fibonacci numbers

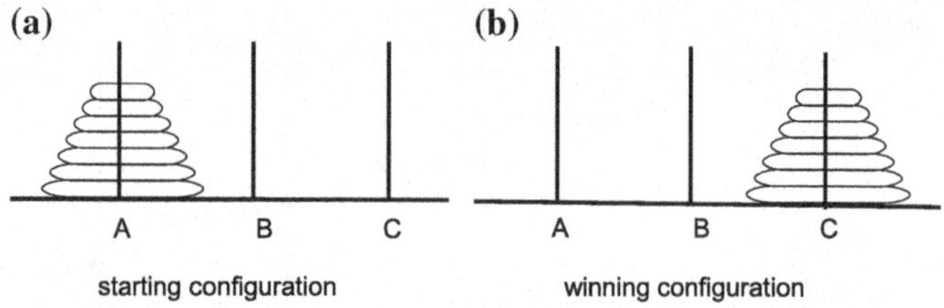

Fig. 9.4 The Tower of Hanoi

4 and 7 are completed, step 3 finishes. After *fib(4)* in the left sub tree is evaluated, the evaluation of *fib(3)* in the right sub tree is evaluated in a similar process, before the return value of *fib(5)*, namely *fib(4)+fib(3)*, can be calculated.

The iterative method in Chap. 5 and the recursive method in this chapter are two different solutions to the same number sequence problem, with different thinkings behind. The computation processes are also different. One problem in the execution of the recursive *fib* calls is that the same Fibonacci number is evaluated multiple times. For example, in Fig. 9.3, *fib(3)* is called twice, *fib(2)* is called 3 times, and *fib(1)* is called 5 times. As a result, this function is less *efficient* than the iterative program in Chap. 5. Efficiency is a very important issue in the study of **algorithms**, or automatic processes to perform a task.

As a final example of problem solving using recursive functions, consider the game of *Tower of Hanoi*. As illustrated in Fig. 9.4, the game consists of three rods and a set of disks of different size, which can be slided onto any of the rods. Denote the three rods as A, B and C, respectively, and the number of disks as n. n can be any positive integer. The game starts with all the disks being placed on rod A, and the goal of the game is to move all the disks onto rod C, under the following rules:

- Each time a disk can be moved from one rod to another;
- At each time, only the top disk on a pile of disks can be moved;
- A disk cannot be placed on top of a smaller disk.

If there is only one disk, a solution is straightforward:

- Move the disk from A to C

If there are two disks, 1 and 2 (from top to bottom), a solution can be:

- Move 1 from A to B
- Move 2 from A to C
- Move 1 from B to C

If there are three disks, 1, 2 and 3 (from top to bottom), a solution can be made by first trying to move the disk 1 and 2 to B, and then 3 to C, and then 1 and 2 to C.

- Move 1 from A to C
- Move 2 from A to B
- Move 1 from C to B

- Move 3 from A to C
- Move 1 from B to A
- Move 2 from B to C
- Move 1 from A to C

In general, if there are n disks, a solution can be made by using the following induction steps.

- First move $n - 1$ disks from A to B
- Then move 1 disk from A to C
- Then move $n - 1$ disks from B to C

Moving *n - 1* disks from *A* to *B* is the same as moving *n - 1* disks from *A* to *C*, except that the target rods differ. As a result, the solution for n disks can be based on the solution for $n - 1$ disks. Based on the thinking above, a recursive function can be defined as follows:

```
>>> def hanoi(n, source, destination, other):
...     if n==1: # basis
...        print 'Move top disk from', source, 'to',
                 destination
...     else: # induction
...        hanoi(n-1, source, other, destination)
...        hanoi(1, source, destination, other)
...        hanoi(n-1, other, destination, source)
...
>>> hanoi(3, 'A', 'C', 'B')
Move top disk from A to C
Move top disk from A to B
Move top disk from C to B
Move top disk from A to C
Move top disk from B to A
Move top disk from B to C
Move top disk from A to C
```

The function *hanoi* above takes four arguments, *n*, *source*, *destination* and *other*, which specify the number of disks, the source rod, the target rod, and the other rod, respectively. It does not use any data structure to model the disks or rods, but simply prints the moving process. The dynamic execution process of the function call *hanoi*(3, '*A*', '*C*', '*B*') is shown in Fig. 9.5. As shown by the examples, recursive functions can be highly declarative, showing the basis and induction steps explicitly.

9.2.2 Functional Programming

As introduced earlier in this chapter, **functional programming** is a programming paradigm that treats computation as the evaluation of functions. In contrast to **imperative programming**, which treats computation as sequences of operations on variables via dynamic control flow, functional programming is *descriptive*, specifying the result without spelling out the steps in which it should be obtained.

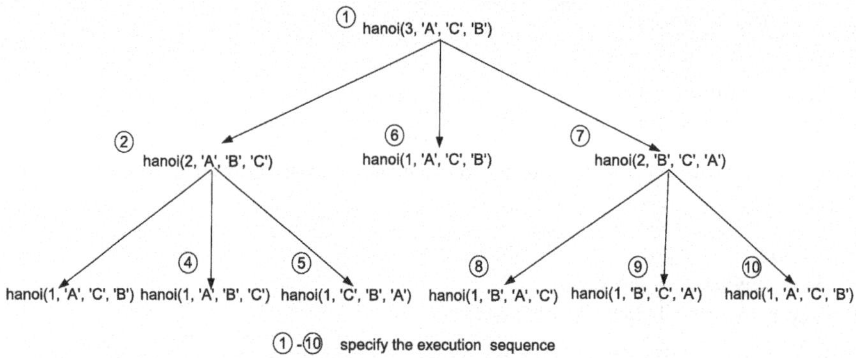

Fig. 9.5 Illustration of recursive function for Tower of Hanoi

Python is an imperative programming language with elements of functional programming. Understanding of functional programming can help writing succinct and more readable programming in Python. Python's support of functional programming include lambda expressions, generator functions, a set of build-in functions such as *map*, *reduce* and *filter*, the ternary operator and list comprehension, some of which have not been introduced in this book. This section begins by introducing the syntax of generator functions, the ternary operator and list comprehension, which are useful not only for functional programming, but also for imperative programming.

Generator functions. A **generator function** is a function that contains one or more **yield** statements. It is special in that the return value is a **generator object**, which can be treated as a *tuple* that allows only iterations, but not *getitem* or *getslice* operations. The object returned by a generator function consists of the sequence of objects, which are determined by the execution sequence of *yield* statements during the execution of the function body. For example:

```
>>> def f():
...     yield 1
...     yield 2
...     yield 3
...
>>> for i in f():
...     print i
...
1
2
3
```

In the example above, the function f contains 3 *yield* statements, which are executed sequentially when $f()$ is evaluated. As a result, the return value of $f()$ is effectively (1, 2, 3). However, being a generator object, the return value can only be iterated over, but not accessed by using the *getitem* or *getslice* operations. Note that the syntax of the *yield* statement is similar to that of the *return* statement. A generator function must return a *None* value.

A function can also contain *yield* statements in dynamic control flow, in which case the dynamic sequence of *yield* statements that are executed determines the returned object.

```
>>> def f():
...     for i in range(1,10):
...         if i % 3 == 1:
...             yield i
...
>>> for i in f():
...     print i
...
1
4
7
```

A generator object records the statements in the corresponding generator function. When it is iterated over, it executes the function body. When a *yield* statement is executed, the current function execution is frozen and the object in the *yield* statement is taken as the current item being enumerated. When the iteration loop asks for the next item, the frozen function execution resumes and runs until an *yield* statement is executed again. The enumeration repeats for each dynamic *yield* statement, until the function execution returns.

The evaluation of a generator object is **lazy** — an item is yielded only when asked for, without the full sequence of items being generated. One advantage of lazy evaluation is the possibility to define an infinite sequence, which is useful to have for functional programming, but cannot be held in finite memory. An example will be given later.

The ternary operator. As introduced in Chap. 2, a **ternary operator** is an operator that requires three operands. There is only one ternary operator in Python, and its syntax is:

```
<expression 1> if <Boolean expression> else <expression 2>
```

The value of the ternary operator is the value of *expression 1* if the Boolean expression is *True*, and the value of *expression 2* otherwise. For example:

```
>>> print 'Yes' if 3 > 5 else 'No'
No
>>> print 'Yes' if 3 <= 5 else 'No'
Yes
```

The ternary operator offers a succinct way of defining a conditional value, which can be used where a single expression is needed, such as for the return value of a *lambda* function. For example, the following function returns the larger of two numbers:

```
>>> larger = lambda x,y: x if x>=y else y
>>> larger(3,5)
5
>>> larger(4,1)
4
```

List comprehension is a succinct way of defining a *list*. Its syntax is:

```
[ <expression> for <variable> in <iterable> if <
    Boolean expression> ]
```

The list comprehension expression above evaluates to a *list* that is equivalent to the code below.

```
l = []    # the equivalent list to be constructed.
for <variable> in <iterable>:
    if <Boolean expression>:
        l.append(<expression>)
```

Here *iterable* is an iterable object, which can be iterated through using a *for* loop. For example:

```
>>> e = [i+1 for i in range(10) if i%3 == 0]
>>> e
[1, 4, 7, 10]
```

The *if* condition can be omitted in a list comprehension expression, which results in all the enumerated items being added to the *list*.

```
>>> e = [i+1 for i in range(10)]
>>> e
[1, 2, 3, 4, 5, 6, 7, 8, 9, 10]
```

Special functions. Python provides a set of built-in functions that are commonly used for functional programming, which include:

- *map(f,x)*, which takes a single-argument function f and an iterable object x as the only two arguments, and returns a *list* that consists of $f(i)$ for each item i in x. For example, if $x = [x_1, x_2, x_3, x_4]$, then $map(f, x)$ returns $[f(x_1), f(x_2), f(x_3), f(x_4)]$.
- *reduce(f,x)*, which takes a two-argument function f and an iterable object x as the only two arguments, and returns a single value by applying f cumulatively to the items in x form left to right. For example, if $x = [x_1, x_2, x_3, x_4]$, the $reduce(f, x)$ returns $f(f(f(1, 2)3)4)$.
- *filter(f,x)*, which takes a single-argument function f and an iterable object x as the only two arguments, and returns a *list* that consists of all the items i in x that satisfies $bool(f(i)) ==$ *True*. f typically returns a Boolean value. When x is a string/*tuple*, the return value of *filter* is a string/*tuple* rather than a *list*.

map, *reduce* and *filter* all *describe* the usage of a function to manipulate a *list*, and are hence useful for functional programming. Both *map* and *filter* can be achieved by using list comprehension. In particular, *map(f,x)* is equivalent to *[f(i)* **for** *i* **in** *x]* and *filter(f,x)* is equivalent to *[i* **for** *i* **in** *x* **if** *f(i)]*.

Functional programming. In addition to the functions *map*, *reduce* and *filter*, the built-in functions *sum*, *min*, *max*, *any* and *all* introduced in Chap. 5 are also useful for functional programming. **Problem solving by functional programming** typically uses *lists* or other iterables to represent the data to be processed, and a set of functions that operate on *lists* to describe the operation. Compared to imperative programming using dynamic control flow, functional programming offers succinct alternative solutions to most basic problems in Chap. 5.

As a first example, consider again the problem of summation. Summing over an iterable collection can be expressed by the *reduce* function:

```
>>> l=[1.1, 2.5, 3.4, -2.6, 0.7]
>>> reduce (lambda x,y:x+y, l)
5.1
```

In the example above, the reduce function applies the + operation to the *list* l, which effectively performs $((((1.1+2.5)+3.4)-2.6)+0.7)$. As shown in Chap. 5, the summation task can also be performed by the function call $sum(l)$. A slight change of the code above gives a functional solution to the factorial problem;

```
>>> def factorial(n):
...     return reduce(lambda x,y:x+y, range(1,n+1))
...
>>> factorial(5)
15
```

As shown in Chap. 5, the method for summation can be generalized into finding the maximum from a container. This problem can also be solved descriptively using the *reduce* function.

```
>>> l=[1.1, 2.5, 3.4, -2.6, 0.7]
>>> reduce(lambda x,y:x if x>=y else y, l)
3.4
```

In the program above, the *lambda* function uses the ternary operator to obtain the larger of x and y. The maximum-finding problem can also be solved by the function call $max(l)$.

Chapter 5 also discussed the problem of deciding whether there is an integer root of $x^4 - 123x - 19260 = 0$ between 1 and 20, showing that this search task can also be formulated as a generalization of summation. A functional solution first maps the range of integers between 1 and 20 to $f(x) = x^4 - 123x - 19260$, and then decides whether any of the results is zero.

```
>>> f=lambda x:x*x*x*x-123*x-19260
>>> reduce(lambda x,y:x or y, map(lambda x:f(x)==0,
                                   range(1,21)))
True
```

In the example above, $map(lambdax : f(x) == 0, range(1, 21))$ results in a *list* of Boolean values, each indicating whether $f(x) = 0$ for an x between 1 and 20. The *reduce* function call achieves $\mathbf{or}_{i=1}^{20}(f(i) == 0)$, the formulation given in Chap. 5. As discussed in the end of Chap. 5, $\mathbf{or}_{i=1}^{20}(f(i) == 0)$ can also be achieved by using the *any* function:

```
>>> f=lambda x:x*x*x*x-123*x-19260
>>> any(map(lambda x:f(x)==0, range(1,21)))
True
```

The counting problem can also be solved descriptively. Chapter 5 discusses the problem of counting the number of integers between 1 and 100 for which the function $sin(x^3)$ is positive. There are two solutions to the problem. The first is to map the *list* of integers between 1 and 100 into a *list* of binary values, where 1 and 0 indicate that the corresponding integer x satisfies $sin(x^3) > 0$ and $sin(x^3) \leq 0$, respectively. Summing over the binary values then gives the answer. The second solution is to filter out all the items such that $sin(x^3) \leq 0$, and the size of the resulting *list* gives the answer.

```
>>> import math
>>> #first solution
>>> sum(map(lambda x: 1 if math.sin(x*x*x)>0 else 0,
                range(1,101)))
60
>>> #second solution
>>> len(filter(lambda x:math.sin(x*x*x)>0,
                    range(1,101)))
60
```

For a more complex example, the problem of induction can be solved by functional programming through the definition of an *infinite list*. Consider the Fibonacci numbers again. The recursive function *fib* earlier in this chapter is a descriptive solution. However, its evaluation leads to much duplicated computation, which is undesired. On the other hand, if the number sequence can be described as an *infinite list*, then it can be evaluated from left to right without duplicated computation. In order to achieve this, one needs to define a function that yields an *infinite list*.

In functional programming, an *infinite list* is typically defined by specifying the **head** and **tail** recursively. Given a *list* l, its head is defined as $l[0]$ and its tail $l[1:]$. A recursive function can be used to define an *infinite list* l by specifying $l[0]$ explicitly, and $l[1:]$ by a call to itself. The key to the definition of such a function is a set of arguments, which is sufficient to specify $l[0]$, and to induce a next set of arguments which can be used for the recursive call for $l[1:]$. For example, the *infinite list* of Fibonacci numbers $[f_0, f_1, f_2 \ldots]$ can be defined as $head_0 = f_0$ and $tail_0 = [f_1, f_2 \ldots]$. Similarly, the tail $[f_1, f_2 \ldots]$ can be defined as $head_1 = f_1$ and $tail_1 = [f_2, f_3 \ldots]$. This definition can repeat to $head_n$ and $tail_n$ infinitely. At each step, knowledge of $head_i$ and $tail_i[0]$ is sufficient for yielding the current f_i and defining $tail_{i+1} = [f_{i+1}, \ldots]$, because $head_{i+1}$ is $tail_{i+1}[0]$ (i.e. f_{i+1}), and f_{i+1} can be derived from $head_i + head_{i-1}$, which is $f_{i-1} + f_i$. As a result, a recursively function $f(head_i, head_{i+1})$, which returns $f_i = head_i$ and $[f_{i+1}, \ldots] = f(head_{i+1}, head_i + head_{i+1})$, is sufficient to describe the Fibonacci numbers. A recursive calling sequence will yield f_i one by one.

For another example, the sequence of factorials $[1!, 2!, 3! \ldots]$ can be described by a function f that takes two inputs i and $n = (i-1)!$, yielding the current head $i! = i \cdot n$ and tail $[(i+1)!, (i+2)! \ldots] = f(i+1, i \cdot n)$.

In general, given a difference equation, an *infinite list* can be defined for a number sequence by using a recursive function f. The key to the definition of f is the set of arguments. The arguments to the function should be sufficient to specify the head of the list, and the tail of the *list* as a call to f itself. A Python implementation of f can take an arbitrary number of arguments, and return a *tuple* $(head, tail)$, where *tail* is a *list* that contains the arguments to the recursive call to f. For example, the function for the Fibonacci numbers can be:

```
>>> ffib=lambda xii,xi:(xi, [xi, xi+xii])
```

xii and *xi* specify f_{i-1} and f_i, respectively. The function for the factorial can be:

```
(continued from above)
>>> ffac=lambda n,i:(i*n, [i*n, i+1])
```

n specifies $(i - 1)!$. Given a recursive function defined in the form above, a generator function can be defined to yield an *infinite list* by a lazy evaluation.

```
(continued from above)
>>> def inflist(f, *init):
...     args = init # arguments to f
...     while True:
...        head, tail = f(*args) # call to get head and tail
...        yield head # yield current item
...        args = tail # prepare next call
```

The follow example shows the use of *inflist* to obtain the first *n* items from an *infinite list*:

```
(continued from above)
>>> n=0
... for i in inflist(ffib, 0, 1):
...     print i
...     if n == 10:
...         break
...     n+=1
1
1
2
3
5
8
13
21
34
55
89

>>> n=0
... for i in inflist(ffac,1,1):
...     print i
...     if n==10:
...         break
...     n+=1
1
2
6
24
120
720
5040
40320
362880
3628800
39916800
```

Note that a *break* statement is necessary to avoid an infinite loop. The code above can be generalized to two functions *getfirstn* and *getnth*, which obtain the first *n* items and the *n*th item form an infinite *list*, respectively.

```
(continued from above)
>>> def getfirstn(f,n): #get first n items from infinite list
...     k=0
...     retval=[]
...     for i in inflist(f,0,1):
...         if k>=n:
...             break
...         retval.append(i)
...         k+=1
...     return retval
...
>>> def getnth(f,n): #get the nth item from infinite list
...     k=0
...     for i in inflist(f,1,1):
...         if k>=n:
...             return i
...         k+=1
...     return None
...
>>> print getfirstn(ffib,10) #first n Fibonacci
[1, 1, 2, 3, 5, 8, 13, 21, 34, 55]
>>> print getnth(ffac, 10) # nth factorial number
39916800
```

As can be seen from the examples, the definition of *inflist*, *getfirstn* and *getnth* allows lazy generation of *infinite lists* by recursive functions. This ends the introduction of functional programming concepts. One final note is that Python is mostly an imperative programming language. For example, Python does not have built-in support for *infinite lists*. There are no built-in functions to perform *inflist*, *getfirstn* or *getnth*. In addition, the functions *map*, *reduce* and *filter* do not yield the output *list* in a lazy mechanism, which makes it possible to use them for *infinite lists* unless custom alternatives are defined. As a result, functional programming can be a nice addition to Python, but in most cases, imperative programming offers the best solution in using Python.

9.3 Files, Serialization and *urllib*

9.3.1 Files

As discussed in Chap. 1, operating systems organize information on the hard disk by files. Each file can be treated a sequence of binary numbers, which represents abstract information. For each type of files, abstract information has its specific interpretation. In text files, for example, binary numbers are interpreted as text characters. In music files, binary numbers can be interpreted as sampled pitches. In image files, binary numbers can be interpreted as the hue and saturation of pixels.

The content of files can be accessed and modified by Python programs. Python provides a convenient interface for reading and writing **text files**. In particular, a **file type** object in Python represents a particular text file on the hard disk. The content of a file can be obtained by calling *reading methods* of a corresponding file object, and changed by calling *writing methods* of a corresponding file object.

As file objects correspond to the physical files, the operating system imposes restriction on their usage, to ensure steady operations of the file system and computer. For example, a file object must be linked to exactly one physical file on the hard disk, and must exist in either a reading mode or a writing mode, but not both. Some files that are critical to the operating system are set as read-only, and cannot be opened in the writing mode.

To open a physical file and link it to a file object, the built-in function *open* can be used. It takes a string argument that specifies the path of the file to be opened, and an optional string argument that specifies a reading/writing mode, returning a file object that is associated with the file. If the second argument is not specified, it is assigned the value 'r' by default, and a file in the reading mode is returned by the *open* call.

For example. Consider the following text file under *Linux* or *Mac OS*:

```
[~/Desktop/abc.txt]
abc
def
ghi
```

A function call to *open('~/Desktop/abc.txt')* returns a corresponding file object in the reading mode. If the specified file does not exist, a call to *open* in the reading mode results in an *IO error*.

Two commonly used file reading methods include *read*, which takes no arguments and returns the content of the corresponding file as a string, and *readlines*, which takes no arguments and returns the content of the corresponding file as a *list* of strings, each containing a line in the file. Line separators are OS-specific. In *Linux* and *Mac OS*, the line separator is the character '\n'; in *Windows OS*, the line separator is the string '\r\n'.

```
>>> file=open('~/Desktop/abc.txt')
>>> type(file)
<type 'file'>
>>> s=file.read()
>>> s
'abc\ndef\nghi\n'
>>> file.close()
>>> file=open('abc.txt')
>>> l=file.readlines()
>>> l
['abc\n', 'def\n', 'ghi\n']
>>> file.close()
```

The example above shows how the content of 'abc.txt' is retrieved by the *read* and *readlines* methods on a corresponding file object, which is bound to the identifier *file*. Note that after the content of the file is read using the *read* method, the *close* method is called to close the physical file, and the *open* method is called a second time to reopen the file before the *readlines* method is called. This is because a file object maintains an index on the physical file, which indicates the location that the next reading operation should start at. The *read* and *readlines* operations move the index to the end of the file, and hence no further reading is possible after a call to them. It is advised to close a file after use, so that the operating system can safely allocate the file to other users.

For a very large file, reading the whole content into an object in memory can fill up the physical memory. Python offers a third reading method, *readline*, which takes no arguments and reads a line from the current location. If the location is the end of the file, an empty string is returned. The following code shows an example use of the method.

```
>>> file=open( '~/Desktop/abc.txt')
>>> s=file.readline()
>>> n=1 # line number
>>> while s:   # if the end of file is not reached,
        there should be at least a '\n' in the line.
...     print 'The', n, 'th line is:', s[:-1] # use
        s[:-1] to strip '\n' in the end
...     n+=1
...     s=file.readline()
...
The 1 th line is: abc
The 2 th line is: def
The 3 th line is: ghi
>>> file.close()
```

The program above prints out the lines of a file by using a *while* loop, incrementally reading each line. A counter *n* is used to record the line index, which is increased by one every time a line is read. Note that the current line *s* is used as the Boolean condition of the *while* loop. When it is empty, the loop terminates. Here *s* is empty only when the end of file is reached. This is because all the lines contain the line separator character '\n' and empty lines in the file yields *s*='\n'. The *while* loop with calls to *readline* is more memory-efficient compared with the *readlines* call. This is reading is performed line by line, and hence the precious line can be recorded from memory by the garbage collector when the next line is processed, thus saving memory.

File objects are iterable, and a *for* loop over a file object enumerates the lines in the corresponding file. It offers a more succinct way of writing the program above.

```
>>> file=open('~/Desktop/abc.txt')
>>> n=1
>>> for line in file:
...     print 'The', n, 'th line is:', line[:-1]
...     n+=1
...
The 1 th line is: abc
The 2 th line is: def
The 3 th line is: ghi
>>> file.close()
```

There are two writing modes in which a file can be opened. The first is 'a', which opens a file for appending new content to its back, and the second is 'w', which opens a file for overwriting its content. The *write* method can be used to write a string to a file in a writing mode. For example,

```
>>> file=open('~/Desktop/abc.txt','a')
>>> file.write('jkl')
>>> file.write('\n')
>>> file.write('mno\npqr')
>>> file.close()
```

The program above modifies *abc.txt* by adding the characters 'jkl\nmno\npqr\n' to its back, which results in the content below.

```
[abc.txt]
abc
def
ghi
jkl
mno
pqr
```

Note that a call to the method *write* does not make a new line implicitly, and new line must be specified by writing a line separator character to the file.

The code below overwrites the content of *abc.txt* entirely.

```
>>> file=open('~/Desktop/abc.txt','w')
>>> file.write('123456')
>>> file.close()
```

Execution of the code above results in *abc.txt* containing the six characters '123456' and no other content. The original content of the file 'abc.txt' is erase immediately when *open* is called. Care should be taken when performing the write operation, so that important content are not lost by careless overwriting of files.

A text file can be used to store structured data for scientific computation. For example, consider the maintenance of examination scores for a class of students. Each student has a unique student ID, and a list of examination scores. In Python, a *dict* can be used to maintain the records, with the keys being student IDs and the values being lists of scores. On the other hand, the records can be stored in a text file, so that they can be kept permanently. Such a file can keep each student in a line, starting with the student ID, followed by a space, and then a space-separated list of scores. The file *score.txt* below shows an example.

```
[scores.txt]
1000101 95 90 70 90 85 65
1000103 99 90 77 85 97 55
1000208 70 35 52 56 60 51
```

Suppose that the file is put into '~/Desktop' on a *Linux* or *MacOS*. It can be loaded into a *dict* object in memory by using the function *load_scores* below.

```
>>> def load_scores(path):
...     d={}
...     file=open(path)
```

```
...     for line in file:           # one student item
...         line=line.split( )      # split into list
...         d[line[0]]=line[1:]     # key=id and value=scores
...     file.close()
...     return d
...
>>> scores=load_scores('~/Desktop/scores.txt')
>>> scores
{'1000208': ['70', '35', '52', '56', '60', '51'],
    '1000103': ['99', '90', '77', '85', '97', '55'],
    '1000101': ['95', '90', '70', '90', '85', '65']}
```

The function *load_scores* takes a string argument that specifies the path of a score file, and returns a *dict* object that contains the content of the record. It works by initializing the return *dict* as an empty *dict*, and then enumerating each line in the file. Each line is split by whitespaces using the *split* method introduced in Chap. 8, and the resulting head and tail are used as the key and value of a new entry in the return *dict*, respectively.

A Python program can be used to obtain a score record from a file, and then edit the record, adding the results of new examinations. For example, the function *save_scores* below can be used to save a score record back into a file.

```
>>> def save_scores(d, path):
...     file=open(path, 'w')
...     for k in d:
...         file.write(k)
...         file.write(' ')
...         file.write(' '.join(d[k]))
...         file.write('\n')
...     file.close()
...
>>> d={'1000101': ['95', '90', '70', '90', '85', '45',
    '70'],'1000103': ['99', '90', '77', '85', '97',
   '55', '71'],'1000208': ['70', '35', '52', '56',
   '60', '51', '40']}
>>> save_scores(d, '~/Desktop/scores.txt')
```

Execution of the program overrides the content of *~/Desktop/scores.txt*.

9.3.2 Serialization Using the pickle Module

The programs above map a *dict* into a string of multiple lines, so that it can be saved into a file and loaded back into memory. The process of saving a memory object into the hard disk as a string is also called **serialization**.

Python provides a module, *pickle*, for automatic serialization of objects. The *pickle* module contains two functions:

- *dump*, which takes an arbitrary object and a file object as the only two arguments, and serializes the object into the file. The method returns *None*.

- *load*, which takes a file object as the only argument and returns an object that has been serialized into the file by *pickle.dump*.

Using the *pickle* module, the loading and saving of score records can be performed in the following way:

```
>>> d={'1000101': ['95', '90', '70', '90', '85',
    '65'],'1000103': ['99', '90', '77', '85', '97',
    '55'],'1000208': ['70', '35', '52', '56', '60',
    '51']}
>>> import pickle
>>> file=open('~/Desktop/scores.txt', 'w')
>>> pickle.dump(d,file)
>>> file.close()
>>> file=open('~/Desktop/scores.txt')
>>> d=pickle.load(file)
>>> d
{'1000103': ['99', '90', '77', '85', '97', '55'],
    '1000208': ['70', '35', '52', '56', '60', '51'],
    '1000101': ['95', '90', '70', '90', '85', '65']}
>>> file.close()
```

9.3.3 Reading Web Pages Using the urllib *Module*

Files on a remote server can be read by using the *urllib* module. Similar to a path name, which identifies a file in a local operating system, a *URL* identifies a file on the Internet.

Most files on the Internet are written in the text form, and typically in HTML, a mark-up language that contain text formatting information, references to media files such as images, and hyperlinks to other files. HTML files can be interpreted by web browsers.

Reading a file from the Internet is similar to reading a file from a local machines. The main difference is that a remote file should be opened by the function *urllib.urlopen*, instead of the function *open*. For example, the example below shows the content of *http://www.scipy.org/index.html*.

```
>>> import urllib
>>> file=urllib.urlopen('http://www.scipy.org/index.
   html')
>>> for line in file:
...     print line[:-1]
...
<!DOCTYPE html>

<html>
  <head>
    <meta charset="utf-8">

    <title>SciPy.org &mdash; SciPy.org</title>
```

```
    <link rel="stylesheet" type="text/css" href=
      "_static/css/spc-bootstrap.css">
    <link rel="stylesheet" type="text/css" href=
      "_static/css/spc-extend.css">
    <link rel="stylesheet" href="_static/scipy.css"
        type="text/css" >
    <link rel="stylesheet" href="_static/pygments.css"
         type="text/css" >
    <link rel="stylesheet" href="_static/scipy-org.css
        " type="text/css" >

    <script type="text/javascript">
      var DOCUMENTATION_OPTIONS = {
        URL_ROOT:     './',
        VERSION:      '',
        COLLAPSE_INDEX: false,
        FILE_SUFFIX: '.html',
        HAS_SOURCE:  true
      };
    </script>
    <script type="text/javascript" src="_static/js/
        jquery.min.js"></script>
    <script type="text/javascript" src="_static/js/
        bootstrap.min.js"></script>
    <script type="text/javascript" src="_static/jquery
        .js"></script>
    <script type="text/javascript" src="_static/
        underscore.js"></script>
    <script type="text/javascript" src="_static/
        doctools.js"></script>
    <link rel="shortcut icon" href="_static/favicon.
        ico">
    <link rel="author" title="About these documents"
        href="about.html" >
    <link rel="top" title="SciPy.org" href="#" >
    <link rel="next" title="Scientific Computing Tools
         for Python" href="about.html" >
  </head>
  <body>
<style type="text/css">
.top-logo-header {
  text-align: left;
  background-color: rgb(140, 170, 230);
  border-bottom: 8px solid rgb(0, 51, 153);
  margin-top: 10px;
  padding: 5px;
  box-shadow: 0px 0px 3px rgb(136, 136, 136);
}
</style>
<div class="container">
  <div class="top-logo-header">
    <a href="#">
      <img style="border: 0;" alt="SciPy" src="_static
          /logo.gif"></a>
  </div>
</div>
```

```
...
  </body>
</html>
>>> file.close()
```

As show by the example, the file object returned by *urllib.urlopen* can be read as local file. In this example, each line of the file is enumerated and printed. The file is in the HTML format, which is beyond the scope of this book. Interested readers are referred to the documentation of *htmllib*, a module provided by Python for handling HTML files.

Exercises

1. Write a function that takes a *list* of *lists* of numbers as the only argument, and returns
   (a) a *list* containing the maximum element in each sub *list*;
   (b) a *list* containing the length of each sub *list*;
   (c) a *list* containing the average of each sub *list*;
   (d) a *list* containing all the items of all the sub *lists*.
2. Write a function that takes a matrix argument and returns
   (a) the number of rows in the matrix;
   (b) the number of columns in the matrix;
   (c) the diagnal of the matrix if it is a square matrix, and *None* otherwise;
   (d) the reverse diagnal of the matrix if it is a square matrix, and *None* otherwise;
   (e) a matrix that consists of all the negated elements;
   (f) the transpose of the input;
   (g) a printable format of the matrix as a string.
3. The built-in *zip* function takes two *list* arguments of the same length, and returns a *list* of *tuples* that contain element-wise pairs from the two *list* argument.

   ```
   >>> l1 = (1, 2, 3)
   >>> l2 = ('a', 'b', 'c')
   >>> zip(l1,l2)
   [(1, 'a'), (2, 'b'), (3, 'c')]
   ```

   Question 6 of Chap. 7 introduces a function *unzip* that takes a *list* of pairs argument, and performs the reverse of *zip*.

   (a) Implement *zip* using *list* comprehension;
   (b) Implement *unzip* using *list* comprehension;
   (c) Implement *unzip* using recursive function calls;
   (d) Implement a function, *zip3(l1, l2, l3)*, which takes 3 *lists* as the arguments and returns a *list* of tuples that consists of element-wise combinations of the arguments;
   (e) Implement a function, *zip(*l)*, which takes an arbitrary number of *lists* as the arguments, and returns a *list* of *tuples* that consists of element-wise combinations of the arguments.

4. Write the following recursive functions
    (a) Write a recursive function, *palindrome_rec*, which takes an integer argument and returns whether it is a palindrome number.
    (b) Write a recursive function, *sum_rec*, which takes a *list* argument, and returns the sum of all the items in the *list*. Assume that the list items are all numbers.
    (c) Write a recursive function, *count_num_rec*, which takes a string argument, and returns the number of numerical characters in the string.
    (d) Write a recursive function, *permutation_rec*, which takes a *list* argument, and prints all possible permutations of its items. Assume that the *list* does not contain duplicate items. For example,

```
>>> permutation_rec([1,2,3])
1 2 3
1 3 2
2 1 3
2 3 1
3 1 2
3 2 1
```

5. Solve the following problem descriptively, using either list comprehension or the functions *map*, *reduce*, *sum*, *any* and *max*.
    (a) find all the powers of 3 in a *list* of numbers;
    (b) find the number of odd numbers in a *list* of numbers;
    (c) decide whether an integer is a palindrome number;
    (hint: turn integer into a *list*, and use Boolean operators)
    (d) decide whether an integer is a prime number;
    (hint: use Boolean operators on a *list* of possible divisors)
    (e) find the longest string in a *list* of strings.

6. Write a function *hanoi_file*, that takes two string arguments that indicate the path of an input and an output file, respectively. The input file contains the number of disks, and the output file should record the moves to solve the puzzle. For example:

```
[input.txt]
3
>>> hanoi_file("input.txt","output.txt")
[output.txt]
Move 1 from A to C
Move 2 from A to B
Move 1 from C to B
Move 3 from A to C
Move 1 from B to A
Move 2 from B to C
Move 1 from A to C
```

Chapter 10
Classes

As introduced in Chap. 2, Python objects belong to types. Various built-in types have been introduced in this book, including *int*, *float*, *long*, *complex*, *str*, *bool*, *tuple*, *NoneType*, *list*, *dict*, *set*, functions and files. Each type supports a specific set of *operators*, such as +, −, *, /, *getitem* (i.e. []), and call (i.e. ()). Some types of objects represent the aggregation of multiple sources of information, rather than a single value. Such types support information integration and access via *attributes* and *methods*. For example, the *complex* type supports the attributes *real* and *imag*, and the method *conjugate*.

```
>>> c=1+2j
>>> d=-1j
>>> c.real
1.0
>>> d.real
0.0
>>> c.imag
2.0
>>> d.imag
-1.0
>>> c.conjugate()
(1-2j)
>>> d.conjugate()
1j
```

Attributes and methods belong to individual objects. As a result, the method calls to *c.conjugate()* and *d.conjugate()* in the example above result in different return values. In addition to methods, operators, such as + and −, also manipulate information about complex numbers. *Operators*, *attributes* and *methods* provides **interfaces** to objects, which are used to access information and perform operations. Python allows custom data structures to be organized with consistent interfaces by defining **custom types**, which follow the same protocols as built-in types, so that code becomes easier to maintain and less prone to bugs.

This chapter discusses **classes**, Python's mechanism for defining custom types, showing how object construction, attributes, methods and operators can be defined for a new type. Classes are the core concept in **object oriented programming**.

Y. Zhang, *An Introduction to Python and Computer Programming*,
Lecture Notes in Electrical Engineering 353, DOI 10.1007/978-981-287-609-6_10

They represent types. In OOP terminology, objects that belong to a type are called **instances** of the type. For example, 1 and −2 are instances of integers, while 'abc' and 'greetings' are instances of strings.

This correlation between built-in types and their objects can be generalized, and used for modeling custom types, namely *classes*. For example, John's bank account is an instance of the account type, and Mary's pet dog is an instance of the dog type. Further, types can have hierarchies. For example, the dog type is a specific sub type of the canine type, which has the wolf type as another sub type. The abstraction of types and instances is central to the concept of object oriented programming, which has fundamental influences on programming languages, making code easier to understand and maintain.

Python supports the most important object oriented programming concepts. For example, the type hierarchy is supported by sub classes in Python. This chapter discusses the basic thinking of object oriented programming, and how it can be applied to Python programming.

10.1 Classes and Instances

Types represents certain types of information. A type can integrate one or more sources of information. This section introduces custom types by using bank accounts as an example. As shown in Table 10.1, a bank account integrates three sources of information, including the account number, the name of the account holder and the balance. Each account instance represents a specific combination of the three attributes.

10.1.1 Classes and Attributes

The **class statement** defines a custom type. Its syntax is shown in Fig. 10.1. A *class* statement starts with the keyword *class*, followed by an identifier, which specifies the name of the new type, or class, and then a semi-colon. After the *class* line is an indented statement block, which typically defines methods, the constructor and operators of the new type.

Table 10.1 Type and instances

Type (account)	Instance 1	Instance 2
Account number	100001	100010
Account holder name	John	Marina
Balance	1000	0

Fig. 10.1 The *class* statement

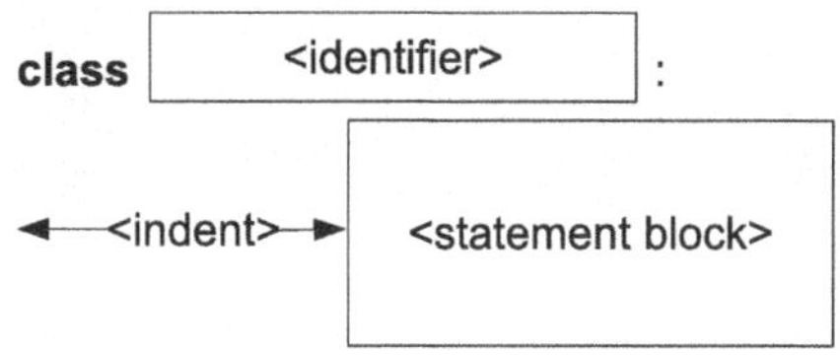

To begin with, consider the simplest case, where a *pass* statement is used as the indented statement block. Such an empty type does not have any methods or operators. Nevertheless, it can be instantiated into objects, which can be assigned attributes. This simplest type of classes shows the basic way in which a custom object integrates various sources of information. Taking bank accounts for example, attributes of an account can include the name of the account holder, the account number and the balance, as shown in Table 10.1. A simplest way of defining the account type is to make it an empty class, assigning relevant attributes to its instances.

```
>>> class Account: # empty type
...     pass
...
>>> a1=Account() # constructor call
>>> a2=Account()
>>> a1.name='John' # attribute assignment
>>> a1.number=100001
>>> a1.balance=1000
>>> a2.name='Marina' # account holder name
>>> a2.number=100010 # account number
>>> a2.balance=0 # balance
>>> a1.balance
1000
>>> a2.name
'Marina'
```

In the example above, the *class* statement defines a new type, and binds it to the identifier *Account* in the global binding table. Two objects, *a1* and *a2*, are constructed, by instantiating the type via the constructor call *Account()*. By this step, neither *a1* nor *a2* hold any attributes.

Then a few **attribute assignment** statements are executed. Attribute assignment is similar to the assignment statement and *setslice* operation, except that the left hand side of the = sign is an attribute, rather than an identifier or slice. The syntax of specifying an attribute in attribute assignment is:

```
<object>.<attribute>
```

This syntax is the same as that for accessing the attribute of an object in an expression (e.g. *c.imag* + 2), which is also referred to as the *getattr* operation. Correspondingly, the attribute assignment statement is also called the *setattr* operation. The correlation between *getattr* and *setattr* is similar to that between *getitem* and *setitem* for container objects, and that between *getslice* and *setslice* for sequential objects.

Similar to the case of *delitem* and *delslice*, the *del* keyword can also be used to perform the *delattr* operation, removing an attribute from an object.

```
(continued from above)
>>> a3=Account()
>>> a3.name='Tim'
>>> a3.name
'Tim'
>>> del a3.name
>>> a3.name
Traceback (most recent call last):
  File "<stdin>", line 1, in <module>
AttributeError: Account instance has no attribute '
   name'
```

As shown by the example, after the attribute *name* is deleted from the object *a3*, a further reference to this attribute leads to an *attribute error*.

Attributes can be arbitrary identifiers, and there is no additional restriction on the name of attributes that can be assigned to objects. The attributes for the account holder's name, for example, can be 'n' rather than 'name'. Attribute assignment can be regarded as attaching variables to objects. Once an attribute is assigned to an object, it is attached to the object and exists throughout the object's existence, unless being explicited deleted by *delattr*.

The assignment of attributes is object-specific rather than type-specific. For example, the attribute 'name' might be assigned to the account object *a1* but not to *a2*, despite that both belong to the *Account* type. To avoid such inconsistencies, it is a standard practice to automatically assign all necessary attributes to an object at its construction, which is achieved by defining a custom **constructor**. As will be shown in the next section, a constructor is a special method, which is executed automatically when new objects are instantiated. As a result, a new object can have an attribute automatically if the attribute is assigned in the constructor.

10.1.2 Methods and Constructors

Methods are special functions defined in the indented statement block in a *class* statement, which apply to all objects that belong to the class. For a simple example, consider defining a method, *deposit*, for the *Account* type. The method takes one argument that specifies the amount to be deposited, and adds the amount to the balance. This can be achieved by defining a special function under the *class* line, as shown below.

```
>>> class Account:
...     def deposit(self, amount):
...         self.balance+=amount
...
>>> a=Account()
>>> a.balance=1000
>>> a.deposit(100) # implicit argument a
```

```
>>> a.balance
1100
>>> Account.deposit(a, 100) # explicit
>>> a.balance
1200
>>> b=Account()
>>> b.deposit(100) # error
Traceback (most recent call last):
  File "<stdin>", line 1, in <module>
  File "<stdin>", line 3, in deposit
AttributeError: Account instance has no attribute '
    balance'
```

The example above defines a class *Account* with a single method, *deposit*, and makes two instances of the class. The method is called on the two instances, and different results are observed. One immediate thing to notice from the definition of *deposit* above is that it takes two arguments rather than one. The second argument, *amount*, is the argument intended for specifying the amount to be deposited. The first argument, *self*, is a special argument necessary for all methods. It specifies the *Account* object of which the *deposit* method is called, and is assigned implicitly in the method call. When the method *a.deposit(100)* is called, *a* is assigned to the argument *self* and 100 is assigned to the argument *amount* before the function body is executed. A call to *a.deposit(100)* is equivalent to a call to *Account.deposit(a,100)*.

Recall the earlier discussion on methods and functions in Chap. 7, where it was mentioned that the role of an object in its method call is similar to that of an argument in a function call. For example, the call to *c.conjugate()* of a complex number *c* would be similar to a call to a global function *conjugate(c)* if such a function existed. The *self* argument in a method definition explains this analogy: *conjugate* is a method defined under a class *complex*, and as all methods do, it takes a *self* argument. When *c.conjugate()* is called, it is translated into a call to *complex.conjugate(c)*, with the object *c* being passed as the *self* argument implicitly.

```
>>> c=1+2j
>>> complex.conjugate(c)
(1-2j)
```

In both *Account.deposit(a, 100)* and *complex.conjugate(c)*, the syntax *<classname>.<methodname>* is used to refer to a method defined in a class. Recall that this is the third use of the dot (.) operator in this book (the first is to refer to an identifier in a module, and the second is to refer to an attribute or method of an object). The underlying correlation between the three uses of the dot operator is discussed later in this chapter. Calls to a method using a class name, with the object being passed explicitly as the first argument (e.g. *Account.deposit(a,100)*), are rarely used, but sometimes necessary, as will be shown in the next section.

In a method call, Python performs **type checking** on the argument assigned to *self*. If it does not belong to the correct class, an error is raised. For example:

```
>>> class A:
...     pass
...
```

```
>>> class B:
...     def f(self):
...         pass
...
>>> a=A()
>>> b=B()
>>> B.f(b)  # okay
>>> B.f(a)  # error
Traceback (most recent call last):
  File "<stdin>", line 1, in <module>
TypeError: unbound method f() must be called with B
   instance as first argument (got A instance instead)
```

The example above leads to an error because the argument *a* in the method call *B.f(a)* does not belong to the type *B*.

Now back to the *Account* example. The function body of *deposit* consists of a single statement, *self.balance+=amount*. When the function body is executed, *self* is assigned to a specific account object *a*. As a result, *a.balance+=amount* is effectively executed. As shown by the example on *Account* above, calling *a.deposit(100)* increases *a.balance* by 100.

In the example, calling *b.deposit(100)* results in an error because the object *b* does not have an attribute named *balance*. When *self.balance+=amount* is executed, *self* is bound to *b*, hence leading to an *attribute error*. As discussed earlier, if all the necessary attributes can be assigned automatically during the construction of account objects, this problem can be solved. A custom constructor serves this goal.

Constructors. As introduced earlier, types integrate information, and the interfaces of a type include attributes, methods, constructors and operators. While attributes, methods and operators have been discussed in this book, **constructors** have not been formally defined. They refer to the mechanism to construct a new object, or to **instantiate** a type. For most built-in types, objects are constructed via literals. For example, the integer literal 123 leads to the construction of a new integer object, and the list literal [1, 2, 3] leads to the construction of a new list object. On the other hand, the construction of a new *set* object is done by calling *set()*. This is a **constructor call**, a function call that takes the name of a type as the function name, and returns a new object of the type. Constructor calls are general and applicable to custom types, although literals are applicable only to the most commonly used built-in types. Similar to other functions, constructors can take arguments. In fact, the type conversion functions, such as *float(3)* and *complex(1, 2)*, can also regarded as constructor calls.

A constructor is a method, but is special in two ways. First, it is called automatically at the evaluation of a constructor call expression, when the class in instantiated. Second, it must be named *__init__*. All the arguments in a constructor call are passed into the corresponding constructor. For example, a constructor of the *Account* class can take three arguments, specifying the account holder name, the account number and the initial balance, respectively, assigning them to corresponding attributes of the new object.

```
>>> class Account:
...     def __init__(self, name, number, balance):
...         self.name=name
...         self.number=number
...         self.balance=balance
...     def deposit(self, amount):
...         self.balance+=amount
...
>>> a=Account('John', 100001, 1000)
>>> a.deposit(100)
>>> a.balance
1100
```

In the example above, the constructor call *Account('John', 100001, 1000)* leads to the constructor *__init__* of the *Account* class being executed, with the new object being assigned to the *self* argument, 'John' being assigned to the *name* argument, 100001 being assigned to the *number* argument and 1000 being assigned to the *amount* argument.

Execution of the function body assigns the three arguments to three attributes of the *self* object, respectively. As a result, the new object *a* has the *balance* attribute once being constructed, and calling the method *a.deposit(100)* immediately after *a*'s construction does not lead to an attribute error. Note that the argument *number* and the attribute *self.number* share the name *number*, but are different variables.

The number of arguments in a constructor call must match the number of arguments in the constructor of the class. For the *Account* class, for example, the constructor takes 3 arguments in addition to *self*, and 3 arguments must be specified in a constructor call. On the other hand, default arguments (Chap. 6), special list arguments (Chap. 7) and special dict arguments (Chap. 8) can also be used in constructors. For example, the default value of the argument *balance* can be made 0, so that it does not need to be specified in a constructor call.

```
>>> class Account:
...     def __init__(self, name, number, balance=0):
...         self.name=name
...         self.number=number
...         self.balance=balance
...     def deposit(self, amount):
...         self.balance+=amount
...
>>> a=Account('John', 100001, 1000)
>>> b=Account('Marina', 100002)
>>> a.balance
1000
>>> b.balance
0
```

In the example, the constructor call *Account('Marina', 100002)* is simpler than the constructor call *Account('John', 100001, 1000)* by using a default value for the third argument *balance*. The constructor can be further simplified by removing the argument *number*, and allocating account numbers automatically. In particular, a solution is to keep an overall account number record, with the starting account number being 100001. Each time a new account is created, the current overall account number is allocated to the new account, and then the overall record is increased by one. This solution can be achieved by using a **class attribute** for the total account number.

10.1.3 Class Attributes and the Execution of a Class Statement

A class attribute can be defined by writing an assignment statement in the indented statement block in a *class* statement. For example:

```
>>> class A:
...     a=1
...
>>> A.a
1
```

In this example, *a* is an attribute to the class *A*, and can be accessed by the expression *A.a*. The use of the dot (i.e. .) operator in *A.a* is similar to the use of the dot operator in the expression *Account.deposit(a, 100)*, which accesses a method of a class. Recall that the *def* statement constructs a new function object, and assigns it to the function name identifier specified in the *def* line. As a result, it can be regarded as a special assignment statement, with the object being assigned being a function object. In this perspective, the definition of class attributes and methods are consistent, both being assignments within a class statement.

Now back to the *Account* type. The overall account number can be assigned to the class *Account* as a class attribute:

```
>>> class Account:
...     total=100001
...     def __init__(self, name, balance=0):
...         self.name=name
...         self.number=Account.total
...         self.balance=balance
...         Account.total+=1
...     def deposit(self, amount):
...         self.balance+=amount
...
>>> a=Account('John') # Account.__init__(a, 'John') called
>>> b=Account('Marina') # Account.__init__(b, 'Marina') called
>>> a.number
100001
>>> b.number
100002
```

As the example above shows, new instances of the *Account* class are given account numbers automatically, starting from 100001. This is achieved by assigning the current value of *Account.total* to the new object, and increasing *Account.total* in the constructor.

In memory, the *class* statement defines a new **class object**, and assigns it to the class name identifier in the *class* line. This is similar to the *def* statement, which defines a function object in memory and gives it a name. However, unlike the *def* statement, the execution of which does not invoke the execution of the indented statement block (i.e. function body), the execution of the *class* statement leads to the execution of the indented statement block under the *class* line. During the execution of the statement block, each time a new identifier is defined, it is taken as a class attribute or method. Similarly, each time a *def* statement is executed, the function is taken as a method.

The example above consists of 5 statements, the first being a *class* statement, the second and third being two assignment statements, with the right hand side of = being constructor call expressions, and the forth and fifth being two *getattr* expressions. When the first statement is executed, a class object is constructed and bound to the identifier *Account* in the global binding table. The indented statement block under the *class* line is executed, with *total* being taken as an attribute of *Account*, and *__init__* and *deposit* as two methods.

When the second statement, *a = Account('John')* is executed, the constructor *Account.__init__* is called automatically, with the argument *self* being assigned the new object, the argument *name* being assigned the string 'John', and the argument *balance* being assigned 0 by default. Execution of the method body assigns *name* to the attribute *self.name*, *Account.total*, which is currently 100001, to the attribute *self.number*, and *balance* to the attribute *self.balance*. Then the value of *Account.total* is increased from 100001 to 100002 by the execution of *Account.total+=1*. The execution of the third statement, *b = Account('Marina')*, follows the same procedure, except that the argument *name* is assigned 'Marina', and the current value of *Account.total* is 100002.

The following example uses a print statement to verify the execution of the *class* statement. The print statement itself does not define attributes or methods, but shows that the indented statement block is executed when a *class* statement is executed.

```
>>> class A:
...     a=1
...     print 'executed'
...     def f():
...         pass

executed
>>>
```

10.1.4 Special Methods

The constructor *__init__* is a special method, called automatically when a new object is constructed. In addition to this method, Python provides a number of other special methods, which are called automatically, and the names of which begin and end with two consecutive underscore characters (i.e. '__'). For example, *__str__* is a special method that is called automatically when string conversion of the object is performed. It must take no arguments (except for *self*) and return a string object that represents the object. For an object *x*, the built-in function call *str*(x) invokes the method call *x.__str__()*, and returns its return value. If the method *__str__* is not defined in the class of *x*, *str*(x) returns a default string representation of *x*. For example:

```
>>> class A:
...     pass # no __str__ method
...
>>> a=A()
>>> str(a) # default string returned
'<__main__.A instance at 0x10a331878>'
>>> print a # invokes str(a) implicitly
<__main__.A instance at 0x10a331878>
>>> class B:
...     def __str__(self):
...         return 'I am a B instance'
...
>>> b=B()
>>> str(b) # custom string returned
'I am a B instance'
>>> print b
I am a B instance
```

Like all methods, the *__str__* method can access the attributes of the object via the *self* argument. The example below shows how a *__str__* method can be defined in the *Account* class, so that information of a specific account can be printed conveniently.

```
>>> class Account:
...     total=100001
...     def __init__(self, name, balance=0):
...         self.name=name
...         self.balance=balance
...         self.number=Account.total
...         Account.total+=1
...     def deposit(self, amount):
...         self.balance+=amount
...     def __str__(self):
...         return """Account holder name: %s\nAccount
      number: %d\nBalance: %.2f""" % (self.name, self.
      number, self.balance)
...
>>> a=Account('John')
>>> b=Account('Marina', 1000)
>>> print a
   Account holder name: John
   Account number: 100001
   Balance: 0.00
>>> print b
   Account holder name: Marina
   Account number: 100002
   Balance: 1000.00
```

In the example, *__str__* accesses attributes through the argument *self*. Note that the order in which methods are defined in a class is not important. When the *class* statement is executed, the methods of the class are defined by the execution of *def* statements, but not executed. The order of method definition in a *class* statement has no correlation with the order of method execution. During the life cycle of an object, the *__init__* method is called first, but then methods can be called in arbitrary orders.

In addition to *__init__* and *__str__* , Python supports a range of special methods which correspond to almost all built-in operators, such as addition, *getitem* and function call, and some commonly-used built-in functions, such as *abs*. The special methods can be roughly classified into type conversion methods, numerical operation methods, numerical comparison methods, collection methods, function methods and attribute methods. *Type conversion methods* are called automatically when the object is converted to a specific type. *__str__* is a type conversion method, called when the object is converted into a string. Other type conversion methods include:

- *__int__(self)*, which is called when the object is converted into *int*, returning an *int* value as the result;
- *__long__(self)*, which is called when the object is converted into *long*, returning a *long* value as the result;
- *__float__(self)*, which is called when the object is converted into *float*, returning a *float* value as the result;
- *__complex__(self)*, which is called when the object is converted into *complex*, returning a *complex* value as the result;
- *__nonzero__(self)*, which is called when the object is converted into *bool*, returning a Boolean value as the result.

Numerical operation methods are called when numerical operators are applied to the object. There are unary and binary numerical operations, which take one and two operands, respectively. *Common numerical operation methods* include:

- *__neg__(self)*, which is called when the unary '−' operator (negate) is applied to the object, returning the corresponding result;
- *__invert__(self)*, which is called when the unary '∼' operator (Boolean bitwise invert) is applied to the object, returning the corresponding result;
- *__abs__(self)*, which is called when the built-in function *abs* is called with the object being the argument, returning the corresponding result.

Binary numerical operation methods include:

- *__add__(self, other)*, which is called when $self + other$ is evaluated, returning the corresponding result;
- *__lshift__(self, other)*, which is called when $self \ll other$ is evaluated, returning the corresponding result;
- *__and__(self, other)*, which is called when *self* & *other* is evaluated, returning the corresponding result.

There are many more binary operators, and their corresponding methods take three general forms, including *__<op>__(self, other)*, *__<rop>__(self, other)* and *__<iop>__(self, other)*, where *op* stands for an operation, such as addition (i.e. *add*). The difference between *__<op>__* and *__<rop>__* is that *__<op>__* is called when the object is the first operand of the binary operator, while *__<rop>__* is called when the object is the second operand. When the expression $x + y$ is evaluated, *x.__ add__(y)* is called automatically. If the method *__add__* does not exist

Table 10.2 Binary numerical methods

<table>
<tr><th>op</th><th>(self, other) operator</th><th>rop(self, other)</th><th>operator</th><th>iop(self, other)</th><th>operator</th></tr>
<tr><td>__add__</td><td>self+other</td><td>__radd__</td><td>other+self</td><td>__iadd__</td><td>self+=other</td></tr>
<tr><td>__sub__</td><td>self-other</td><td>__rsub__</td><td>other-self</td><td>__isub__</td><td>self-=other</td></tr>
<tr><td>__mul__</td><td>self*other</td><td>__rmul__</td><td>other*self</td><td>__imul__</td><td>self*=other</td></tr>
<tr><td>__div__</td><td>self/other</td><td>__rdiv__</td><td>other/self</td><td>__idiv__</td><td>self/=other</td></tr>
<tr><td>__floordiv__</td><td>self//other</td><td>__rfloordiv__</td><td>other//self</td><td>__ifloordiv__</td><td>self//=other</td></tr>
<tr><td>__mod__</td><td>self%other</td><td>__rmod__</td><td>other%self</td><td>__imod__</td><td>self%=other</td></tr>
<tr><td>__pow__</td><td>self**other</td><td>__rpow__</td><td>other**self</td><td>__ipow__</td><td>self**=other</td></tr>
<tr><td>__lshift__</td><td>self<<other</td><td>__rlshift__</td><td>other<<self</td><td>__ilshift__</td><td>self<<=other</td></tr>
<tr><td>__rshift__</td><td>self>>other</td><td>__rrshift__</td><td>other>>self</td><td>__irshift__</td><td>self>>=other</td></tr>
<tr><td>__and__</td><td>self&other</td><td>__rand__</td><td>other&self</td><td>__iand__</td><td>self&=other</td></tr>
<tr><td>__or__</td><td>self|other</td><td>__ror__</td><td>other|self</td><td>__ior__</td><td>self|=other</td></tr>
<tr><td>__xor__</td><td>self∧other</td><td>__rxor__</td><td>other∧self</td><td>__ixor__</td><td>self∧=other</td></tr>
</table>

for *x*, then *y.__radd__(x)* is called automatically. If *__radd__* does not exist for *y* either, an *unimplemented error* is raised.

The *__<iop>__* methods are called for augmented arithmetic assignment operators, modifying the object in place. For example, *x.__iadd__(y)* is called when *x+=y* is evaluated. A full list of commonly used binary numerical methods are shown in Table 10.2.

Comparison Methods. The most commonly used numerical comparison method is *__cmp__(self, other)*, which is called when *self>other*, *self>=other*, *self<other*, *self<=other*, *self==other*, *self!=other* or the built-in function call *cmp(self, other)* is evaluated. The return value of *cmp(self, other)* and *self.__cmp__(other)* should be positive if *self>other* is true, zero if *self==other* is true, and negative if *self<other* is true.

For more fine-grained comparison, the methods *__gt__(self, other)*, *__ge__(self, other)*, *__lt__(self, other)*, *__le__(self, other)*, *__eq__(self, other)* and *__ne__ (self, other)* are called when *self>other*, *self>=other*, *self<other*, *self<=other*, *self==other*, *self!=other* are evaluated, respectively. All the methods return a Boolean value, indicating whether the corresponding comparison is true. The methods have higher priority compared to the *__cmp__(self, other)* method. For example, when the *__ge__(self, other)* is defined, it is called instead of *__cmp__(self, other)* when *self>other* is evaluated.

Commonly used **container methods** include:

- *__len__(self)*, which is called when the built-in function *len* is called, with the object being the argument, and returns the size of the object. If the Boolean conversion of the object is requested but *__nonzero__(self)* is undefined, then *__len__(self)* is called, and the Boolean conversion is true if and only if the return value of *__len__(self)* is non-zero.

- *__contains__(self, item)*, which is called when the expressions *item in self* and *item not in self* are evaluated, returning a Boolean value, which is *True* if and only if *item* is a member of *self*.
- *__getitem__(self, key)*, which is called when the *getitem* operation is applied, returning the corresponding item;
- *__setitem__(self, key, value)*, which is called when the *setitem* operation is applied;
- *__delitem__(self, key)*, which is called when the *delitem* operation is applied;
- *__getslice__(self, b, e)*, which is called when the *getslice* operation *self[b:e]* is executed, returning an object that has the same type as *self*;
- *__setslice__(self, b, e, slice)*, which is called when the *setslice* operation *self[b:e]= slice* is executed;
- *__delslice__(self, b, e)*, which is called when the *delslice* operation *del self[b:e]* is executed.

Function Method. The only *function method* is *__call__(self, [,args])*, which is called when the object is called as a function. For example, when the function call expression *x(arg1, arg2, arg3)* is evaluated, *x.__call__(arg1, arg2, arg3)* is called automatically, with the return value being used for the return value of the *x* function call.

Attribute Method. There are three *attribute methods*, including

- *__getattr__(self,name)*, which is called when *self.name* is evaluated, but *name* is not an attribute of *self*, returning the custom attribute value;
- *__setattr__(self,name,value)*, which is called when *self.name=value* is executed, performing custom attribute setting.
- *__delattr__(self,name)*, which is called when *del self.name* is executed, performing custom attribute deletion.

10.1.5 Class Examples

Classes can be used to organize data structures, making them easier to maintain. In the *Account* example, various sources of information about an account are aggregated to a single *Account* object, and managed by attributes and methods. Without a custom type definition by the *class* statement, it would be more difficult to write a program that manages such information.

For another example use of classes, consider the stack data structure introduced in Chap. 7. A stack is a sequential data structure that manages its items in the last-in-first-out order. In Chap. 7, a stack is represented by a list, and stack operations are implemented using global functions. A disadvantage of this approach is the separation of data and operations. Given a list, there is not an explicit way of telling whether it represents a stack or not. This issue can be addressed by defining a class for the stack data structure, which offers centralized organization of data and operations.

In particular, a *Stack* instance can use a list attribute to hold its items, and provide the pushing and popping operations as methods.

```
>>> class Stack:
...     def __init__(self, l=[]):
...         self.l=l[:]
...     def push(self, x):
...         self.l.append(x)
...     def pop(self):
...         return self.l.pop(-1)
...     def __len__(self):
...         return len(self.l)
...     def __str__(self):
...         return 'Stack:'+' '.join([str(e) for e in self.l])
...
>>> s1=Stack() # an empty stack
>>> s2=Stack([1, 2, 3]) #Bottom 1 2 3 Top
>>> s2.pop() # BOTTOM 1 2
3
>>> s2.push(0) # BOTTOM 1 2 0
>>> s1.push(5) # BOTTOM 5
>>> print s1
Stack:5
>>> print s2
Stack:1 2 0
>>> s2.pop() # BOTTOM 1 2
0
>>> s1.pop() # BOTTOM
5
>>> s2.pop() # 1
2
>>> if s1: # __len__ is called
...     s1.pop() # not executed
...
>>>
```

The class *Stack* above contains three special methods. The first is *__init__*, the constructor, which allows a *Stack* object to be initialized as an empty stack, or from a *list* that specifies the initial items. The use of *Stack()* and *Stack([1, 2, 3])* to construct *Stack* objects is similar to the use of *set()* and *set([1, 2, 3])* to construct *set* objects. In the constructor, the argument *l* is copied before being assigned to *self.l*, so that the attribute *self.l* is bound to a different object, and further changes to *l* do not affect *self.l*. Since the attribute *self.l* is assigned in the constructor, all *Stack* objects have this attribute.

The second special method is *__len__*, which is called automatically when the functions *len* and *bool* are called. The size of *self.l* is returned by this method, indicating the number of items on the stack. The third special method is *__str__*, which is called automatically when the stack is converted into a string.

As shown by the example, the use of a class allows stack objects to hide the internal *list* representation from users, with all operations being achieved by method interfaces. The hiding of implementation details frees the user of stacks from unnecessary concerns, and makes reimplementation of the *Stack* type using other internal representations easy. This advantage is called **encapsulation**.

Similar to stacks, matrices introduced in Chap. 9 can also be defined as a custom type.

```
>>> class Matrix:
```

```
...     def __init__(self, l=[]):
...         import copy
...         self.l=copy.deepcopy(l)
...     def get_row(self, i):
...         return self.l[i][:]
...     def get_column(self, i):
...         s=[]
...         for row in self.l:
...             s.append(row[i])
...         return s
...     def __mul__(self, a):
...         import copy
...         if type(a)==int:
...             N=copy.deepcopy(self.l)
...             for i in range(len(N)):
...                 for j in range(len(N[i])):
...                     N[i][j]*=a
...             return Matrix(N)
...     def __add__(self, other):
...         import copy
...         R=copy.deepcopy(self.l)
...         S=copy.deepcopy(other.l)
...         for i in range(len(R)):
...             for j in range(len(R[i])):
...                 R[i][j]+=S[i][j]
...         return Matrix(R)
...     def __str__(self):
...         s=''
...         for i in range(len(self.l)):
...             for j in range(len(self.l[i])):
...                 s+=str(self.l[i][j])
...                 s+='\t'
...             s+='\n'
...         return s
...
>>> m1=Matrix([[1, 2, 3], [4, 5, 6], [7, 8, 9]])
>>> m2=Matrix([[1, 0, 0], [0, 1, 0], [0, 0, 1]])
>>> print m1
1    2    3
4    5    6
7    8    9

>>> print m2
1    0    0
0    1    0
0    0    1

>>> m3=m2*-1
>>> print m3
-1   0    0
0    -1   0
0    0    -1

>>> m4=m1+m3
>>> print m4
0    2    3
```

```
4    4    6
7    8    8
```

The class *Matrix* above implements some matrix operations discussed in Chap. 9, including *get_row*, *get_column*, scaling and addition. It provides a centralized organization of data, with encapsulation, and is therefore easier to use compared with a set of global functions that directly handle *lists* of *lists*, as given in Chap. 9.

The matrix addition and scaling operations are implemented using the special methods *__add__* and *__mul__*, respectively. In *__mul__*, type checking is performed to decide whether the second operand is an integer or not. The identifier *int* here represents the *int* type. Type names and constructors having been introduced, it is easy to draw the correlation between *int* and *int(3, 5)* by considering that between *Stack* and *Stack([1, 2, 3])*, and that between *set* and *set([1, 2, 3])*. If the second operand to *__mul__* is an integer, scaling is performed; otherwise the method performs nothing and returns *None*. An exercise problem asks to complete the *__mul__* method by also performing matrix multiplication.

10.1.6 The Underlying Mechanism of Classes and Instances

As discussed earlier, the dot (.) operator has three main uses. The first is to access an identifier from a module, the second is to access an attribute from an object, and the third is to access a class attribute. This suggests some underlying connection between modules, objects and classes. Recall that an object can be assigned arbitrary attributes, and the *class* statement allows arbitrary attribute and method. Python supports such freedom by associating class and instance objects with binding tables. The *setattr* operation simply adds a new identifier to the binding table of the corresponding object. Similarly, when a *class* statement is executed, all the identifiers defined in the indented statement block (e.g. in assignment statements, and *def* statements) are inserted into the binding table of the class object.

A class is a Python object in memory. The uniqueness of class objects is that they can be instantiated. The binding tables between a class object and objects of its instances are hierarchical, where the class binding table can be viewed as the a *outer-scope* binding table of an instance binding table. When Python looks for an attribute in an instance but cannot find it, it goes to the class binding table to search for it. For example,

```
>>> class A:
...     x=1
...
>>> a=A()
>>> b=A()
>>> a.x
1
>>> b.x
1
>>> A.x+=1
```

```
>>> a.x
2
>>> b.x
2
```

In the example above, *a* and *b* are two instances of *A*, which has an attribute *x*. Although neither *a* nor *b* has an attribute named *x*, *a.x* and *b.x* can be accessed, which refer to *A.x*. When *A.x* changes, both *a.x* and *b.x* change, because they refer to *x* in *A*'s binding table.

Assignment of instance attributes, on the other hand, is always carried out in the instance binding table, but never in the class binding table. For example:

```
(continued from above)
>>> a.x=3
>>> A.x
2
>>> a.x
3
>>> b.x
2
>>> b.x-=1
>>> a.x
3
>>> b.x
1
>>> A.x
2
```

In the example above, when *a.x=3* is performed, a new identifier *x* is added to *a*'s binding table. Therefore, *a.x* becomes different from *b.x*. When *b.x*−= 1 is executed, *b.x* =*b.x*−1 is executed, with the right hand of the = sign being evaluated first, and then assigned to the attribute on the left hand side of =. During the evaluation of *b.x*−1, since the identifier *x* is not found in *b*'s binding table, Python goes to *A*'s binding table to find *x*, which is 2. The result of 2 − 1 is in turn assigned to a new identifier *x* in *b*'s binding table, which makes *b.x* different from *A.x*.

To illustrate the underlying mechanism for the *class* statement and new object instantiation, consider again the *Account* class. Figure 10.2 shows the memory structure when the class definition of *Account* and the instantiation of *a=Account('John')* are executed. As shown by Fig. 10.2a, when the *class* statement is executed, a new binding table is constructed for the class object, and the indented statement block is executed. During the execution, new objects are constructed, and bound to identifiers in the binding table of the new class object. Note that methods have their *outer-scope links* pointing to the global binding table, rather than the class binding table.

The instantiation of a class object consists of two steps: the construction of the new instance object and the calling of the constructor, as shown in Fig. 10.2b, c, respectively. When the new instance object is constructed, a corresponding binding table is constructed. The new binding table has an outer-scope link to the class binding table, for hierarchical look-up of identifiers. When the constructor call is executed, a function binding table is constructed for the local scope of the constructor, as introduced in Chap. 6. As indicated by the outer-scope pointer in the function object,

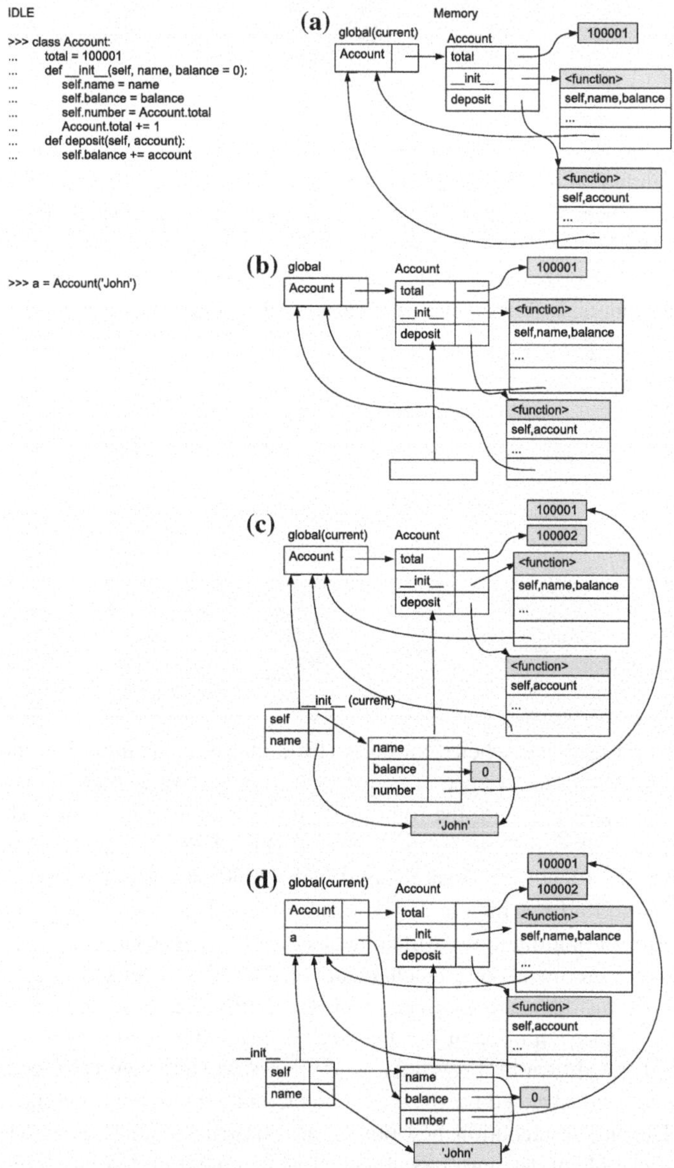

Fig. 10.2 Memory structures of class definition and instantiation. **a** The *class* statement. **b** New object constructed. **c** Constructor called. **d** Assignment done

the outer-scope of this binding table is the globe binding table, rather than the class binding table. When the statement *self.number = Account.number* is executed, the identifier *self* is found in the local binding table, but the identifier *Account* is not. Python follows the outer-scope link to find *Account* in the global binding table.

During the execution of the assignment statement *a=Account('John')*, the right hand side of = is evaluated first, leading to an instantiation of the class *Account*. The value of the expression *Account('John')*, which is the new instance object, is assigned to the identifier *a* in the global binding table, in which this statement is executed, as shown in Fig. 10.2d.

10.2 Inheritance and Object Oriented Programming

Types have hierarchies. For example, both dogs and wolfs are sub categories of the canine category. Since all canines have a head, four legs and a tail, definition of these attributes would be unnecessary for dogs and wolfs given that they are sub categories of the canines. The concept of **class extension** comes from this observation. If a class describes a specific sub type of a general type, and a slight modification of the general type is sufficient for specifying the sub type, then **inheriting** attributes and methods from the general class and writing only the difference in the new class will reduce code duplication.

To facilitate description, a final version of the *Account* class can be saved into a file.

```
[account.py]
class Account:
    total=100001
    def __init__(self, name, balance=0):
        self.name=name
        self.balance=balance
        self.number=Account.total
        Account.total+=1
    def deposit(self, amount):
        if amount<=0:
            print 'Error, negative amount.'
            return
        self.balance+=amount
    def withdraw(self, amount):
        if amount<0:
            print 'Error: negative amount'
            return
        if amount>self.balance:
            print 'Error: insufficient fund'
            return
        self.balance-=amount
    def __str__(self):
        return '''Account holder name: %s
Account number: %d
Balance: %.2f ''' %(self.name, self.number, self.
   balance)
```

Two changes are made to the *Account* class in this version, compared to the version in the previous section. First, the *deposit* method now checks whether the amount is negative, in which case an error is reported and the deposit action is cancelled.

Second, a *withdraw* method is added to the class, which takes a single argument that specifies an amount, and deduces it from the balance. The *withdraw* method checks both whether the input amount is negative and whether it is larger than the current balance; in both cases errors are reported and the withdraw action is cancelled.

10.2.1 Sub Classes

In Python, the type hierarchy is achieved by **sub classes**. A sub class is an **extention** to a **base class**, which represents a specific sub type of the base class in a type hierarchy. For example, the wolf type can be represented by a sub class of a canine type, if these types are modeled in Python.

As mentioned earlier, sub classes can make Python programs simpler by inheriting attributes and methods of the base class, so that code duplication is minimized. To give an example, consider a *CreditAccount* class, which represents a specific type of accounts that has a credit limit. Overdraft of funds is allowed within the credit limit, resulting in a negative balance. In all the other aspects, a credit account behaves exactly the same as the base *Account* class.

The *CreditAccount* class can be developed in two steps. First, an empty extention of the base class *Account* can be made, which inherits all the attributes and methods of the *Account* class. Second, the credit limit feature can be added by overriding some existing functionalities.

The syntax for defining a sub class of an existing class is shown in Fig. 10.3. It is an extension to the *class* statement, with the only difference being an additional base class name after the class name, which is enclosed in a pair of round brackets. Using this statement, the base class attributes and methods are inherited automatically in the sub class. The statement block under the *class* line defines the attributes and methods that are unique in the sub class.

To make an empty extension to the class *Account*, a single *pass* statement can be used as the statement block of the new class. An example is shown below.

```
>>> from account import Account
>>> class CreditAccount(Account):
...     pass
...
>>> a1=Account('John')
>>> a2=CreditAccount('Marina')
>>> a3=Account('Catherine')
```

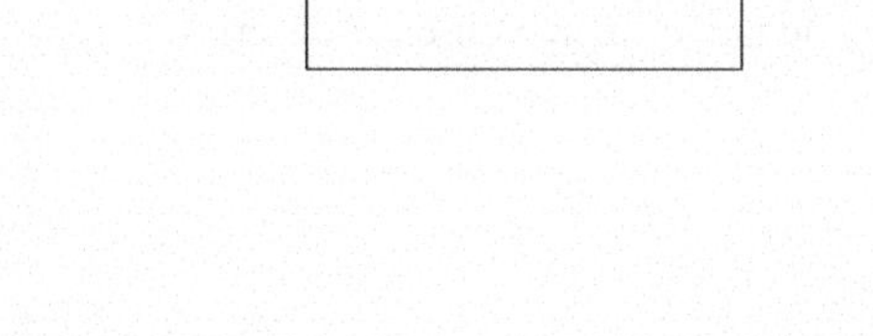

Fig. 10.3 The *class* statement with class extension

```
>>> print a1
Account holder name: John
Account number: 100001
Balance: 0.00
>>> print a2
Account holder name: Marina
Account number: 100002
Balance: 0.00
>>> print a3
Account holder name: Catherine
Account number: 100003
Balance: 0.00
>>> a2.deposit(100)
>>> Account.deposit(a2, 100) # a2 is Account
>>> CreditAccount.deposit(a2, 100) # a2 is
    CreditAccount
>>> CreditAccount.deposit(a1, 100) # a1 is not
    CreditAccount
Traceback (most recent call last):
  File "<stdin>", line 1, in <module>
TypeError: unbound method deposit() must be called
    with CreditAccount instance as first argument (got
    Account instance instead)
```

In the example above, the class *CreditAccount* is defined as empty extension to the class *Account*. Several instances of *Account* and *CreditAccount* are constructed. It can be seen that the *CreditAccount* instance *a2* has all the attributes and methods of *Account* instances, and behaves almost identically compared with the *Account* instances *a1* and *a3*. For example, the constructor of *a2* is the same as that of *a1* and *a3*, taking a *name* argument. The methods *__str__* and *deposit* also apply to *a2*. Interestingly, *a2* also participates in the automatic allocation of account numbers.

The reason behind these observations is that, when a *CreditAccount* method instance is called, but the method is not defined in the *CreditAccount* class, the corresponding method in *Account* is called. For example, when *a2.deposit(100)* is called, Python tries to call *CreditAccount.deposit(a2, 100)*. However, *deposit* is not defined in *CreditAccount*. As a result, Python calls the base class method *Account.deposit(a2, 100)*. The same happens for *__init__* and *__str__*. When *a2=CreditAccount('Marina')* is executed, *Account.__init__(a2, 'Marina')* is called.

As mentioned earlier, Python performs type checking when the *self* argument is assigned. Apparently, the *CreditAccount* instance *a2* passes the type checking of *Account* methods, which implies that *a2* is also an *Account* instance. This is intuitively understandable: a dog instance is also a canine instance, for example. On the other hand, a canine instance is not necessarily a dog instance, and an *Account* instance is not necessarily a *CreditAccount* instance. Hence, when passing *a1* to *CreditAccount.deposit*, Python reports an error, despite that *CreditAccount.deposit* leads to *Account.deposit*.

10.2.2 Overriding Methods

Having inherited all the *Account* methods, the next step to take for the definition of *CreditAccount* is to define the parts that are different from the base *Account* class. There should be two main changes. First, the constructor needs to be modified, adding the assignment of another attribute: the credit limit. Second, the *withdraw* method needs to be modified, so that overdraft is allowed within the credit limit.

To apply the changes above, the constructor and the *withdraw* method need to be redefined in the new class, **overriding** those in the base class.

```
>>> from account import Account
>>> class CreditAccount(Account):
...     def __init__(self, name, balance=0, limit=0):
...         Account.__init__(self, name, balance)
...         self.limit=limit
...     def withdraw(self, amount):
...         if amount<0:
...             print 'Error: negative amount.'
...             return
...         if amount>self.balance+self.limit:
...             print 'Error: insufficient fund'
...             return
...         self.balance-=amount
...
>>> a1=Account('John')
>>> a2=CreditAccount('Marina', limit=500)
>>> a3=Account('Catherine', 1000)
>>> a1.balance
0
>>> a2.balance
0
>>> a3.balance
1000
>>> a1.withdraw(100)
Error: insufficient fund
>>> a2.withdraw(100)
>>> a3.withdraw(100)
>>> a1.balance
0
>>> a2.balance
-100
>>> a3.balance
900
```

In the example above, *CreditAccount* is redefined, with its own constructor and *withdraw* method. Several instances of *Account* and *CreditAccount* are constructed, which now behave diffferently. The *CreditAccount* instance, *a2* is constructed by specifying a credit limit, *limit*=500. Initially, the balances of both *a2* and the *Account* instance *a1* are 0s. When *a1.withdraw(100)* is called, an error is reported and the transaction is cancelled. However, when *a2.withdraw(100)* is called, the transaction proceeds even though the balance is smaller than the amount withdrawn, leaving a negative balance in *a2*. The different behavior results in the overriding of the *withdraw* methods.

The new *CreditAccount* instances have a new attributes, *limit*, which specifies the credit limit of the account. The constructor of *CreditAccount* takes an additional argument that specifies the credit limit, and assigns it to the *limit* attribute of the new instance to be constructed. The rest of the construction, including the assignment of *name*, *number* and *balance*, is the same as that for *Account*. As a result, the constructor of *CreditAccount* simply calls the base constructor *Account.__init__*, with the new *CreditAccount* instance *self* being passed as the *self* argument, and the input *name* and *balance* arguments being passed as the *name* and *balance* arguments for *Account.__init__*, respectively. Execution of *Account.__init__* in turn assigns the respective attributes to *self* the new *CreditAccount* instance, and increases *Account.total* by one.

The *withdraw* method of *CreditAccount* is different from that of *Account* in that the prerequisite is *amount<balance+limit* now instead of *amount<balance*. Because the new prerequisite is less strict than the prerequisite in the base *withdraw* method, the new *withdraw* method cannot call the base method in the same way as the *__init__* method does, but has to rewrite the complete procedure. Nevertheless, by inheriting from the base class, the *CreditAccount* class is significantly simplified as compared to an implementation written from scratch.

10.2.3 The Underlying Mechanism of Class Extention

Python's mechanism for sub class is rather simple. The binding table of a class that has a base class has an **outer-scope link** that points to the binding table of the base class, and Python follows the link to find identifiers that cannot be found in the class. For example, the memory structure of the previous example, after the three instances *a1*, *a2* and *a3* are constructed, is shown in Fig. 10.4, where the outer-scope link of *CreditAccount* is *Account*.

As an example that illustrates how Python uses the memory structures, consider the following process. When *a2=CreditAccount('Marina', limit=500)* is called, the constructor *CreditAccount.__init__* is executed, with a new local binding table being constructed for the function. In the local binding table, there are four entries: *self*, which points to the new instance, *name*, which points to the string object 'Marina', *balance*, which points to the *int* object 0, and *limit*, which points to the *int* object 500. The outer-scope link of the local binding table for *CreditAccount.__init__* is the global binding table. When the first statement in the function body, *Account.__init__(self, name, balance)*, is called, Python looks for *Account* in the local binding table. However, it does not exist here. Thus Python follows the outer-scope link to find it in the global binding table, which points to the binding table of *Account*, where *__init__* is found.

A call to the function *Account.__init__* leads to the construction of its own local binding table, which contains three entries: *self*, which points to the same new

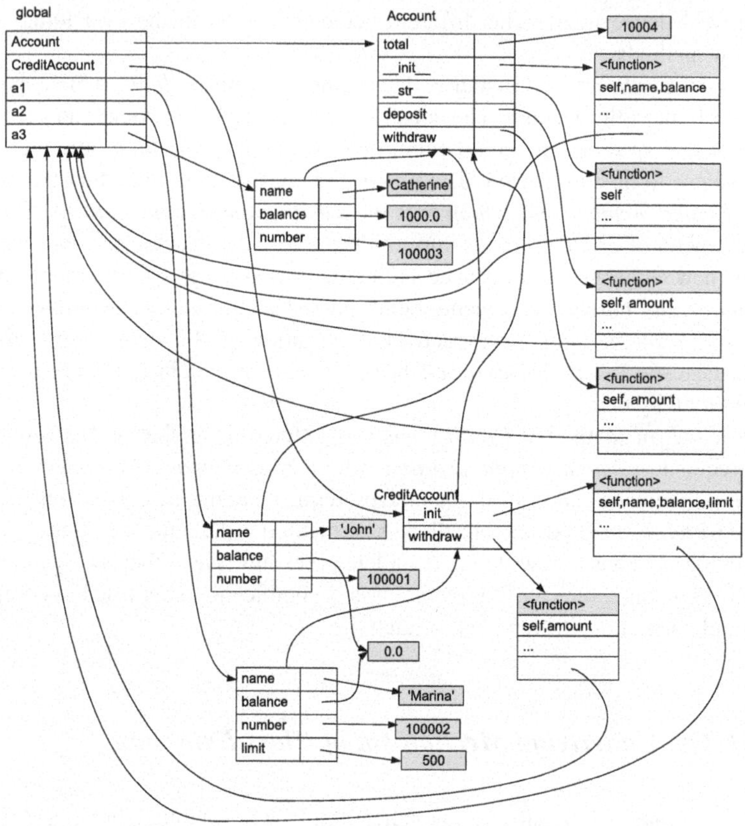

Fig. 10.4 Memory structures of sub class

instance, *name*, which points to the *str* object 'Manina', and *balance*, which points to the same *int* object 0. The binding table has an outer-scope link that points to the global binding table, as in the case of Fig. 10.4 (c). Execution of the function body results in the attributes *name*, *number* and *balance* being created in the binding table of *self*. After the call to *Account.__init__* returns, the second statement in *CreditAccount.__init__* is executed, and the attribute *limit* is created in the binding table of *self*. At this point, the call to *CreditAccount.__init__* finishes, and the *self* object is assigned to the identifier *a2* in the global binding table.

An exercise problem asks for the memory structures when the process above is executed. The changes in each step is mechanical, following the execution rules of each statement. For example, the mechanism of function calls and assignments have been discussed in Chaps. 6 and 2, respectively. Here in this process, they are applied to class binding tables, but the mechanisms remain the same.

10.2.4 Object Oriented Programming

Functions and classes are two of the most influential concepts in programming. Central to **procedural programming**, functions allow code to be modularized into sub procedures, so that complex functionalities are represented by simple function calls. Functions are so influential to computer programming that it is typically supported by hardware also.

Classes, on the other hand, offers modularized organization of both code and data. The concept of classes is central to **object oriented programming**, which has had a significant impact on programming languages over the past few decades, including C++, Java and Python. The two inventors of object oriented programming both won the Turing award, the most esteemed honor in the field of computer science.

There are three most important elements in object oriented programming, namely **encapsulation**, **inheritance** and **polymorphism**. **Encapsulation** refers to the aggregation of attributes and methods in a class, which represents a complex data type, and the hiding of implementation details by using methods as interfaces to operations on instances. **Inheritance** refers to the availability of base class attributes and methods in sub classes, which enables code reuse in a type hierarchy. **Polymorphism** refers to different behaviors of the same method between different objects. For example, when the '+' operator is applied to two numbers, it calculates the sum of the two operands; but when applied to two sequence objects, such as strings, it calculates the concatenation of the two operands.

Another typical example of polymorphism is the shape hierarchy in a graphical user interface. Suppose that the *shape* type is a general type, and the *square*, *circle* and *triangle* types are its sub types. Each of the three sub types has a method, *draw*, which draws itself on the monitor. The behavior of *draw* is different when the sub type is different, drawing a different shape. However, as long as the user knows that an object is an instance of the *shape* type, she can call the *draw* method of the object, drawing it on the monitor. The drawing behavior is polymorphic.

It is a good habit to put frequently used sub routines into functions, and the most important data structures into classes. However, there is no simple formula for the best design of function and class interfaces, and a better design can often lead to significantly less cost in code maintenance, particularly when the software program becomes large. How to design the most readable code is an important subject in **software engineering**, and it takes practice to build experience on this skill.

10.3 Exception Handling

Many types of errors have been touched upon in this book, including *value errors* in Chaps. 2, 3, 7 and 8, *index errors* in Chaps. 3 and 8, *type errors* in Chaps. 3, 6, 7, 8 and 10, *name errors* in Chaps. 3 and 6, *attribute errors* in Chaps. 3, 6 and 10, *syntax errors* in Chap. 4, *unbounded local errors* in Chap. 6, *key errors* in Chap. 8, and *runtime errors* in Chap. 8. All the errors are represented by camel case words in the console output. For example, a value error is repeated with a line that begins with the word 'ValueError', followed by some details about the error. Similarly, a name error report begins with the word 'NameError'.

The camel cased words are type names. When an error happens during the execution of code, Python makes an instance of the corresponding error type, and use it to store information about the error that has occurred. The error object is also called an **exception**. There is a large set of exception types in Python, and they form a type hierarchy. As shown in Fig. 10.5, all exception types are direct or indirect sub types of a *BaseException* type. Most of the exception types in this book are sub types of the *StandardError* type. For example, both *IndexError* and *KeyError* are direct sub types of *LookupError*, which is a sub type of *StandardError*. *StandardError* is a sub type of *Exception*. Besides *Exception*, another sub type of *BaseException* is *KeyboardInterrupt*, which has been seen in Chaps. 4 and 9, recording a keyboard interrupt of program execution.

All the aforementioned error types are available in the *__builtins__* scope. The type hierarchy of exceptions makes it easier to understand the relationship between different exceptions. In addition, it also makes it easier for an **exception handler** to deal with different types of exceptions with deferent procedures. The remainder of this section introduces the dynamic control flow for exceptions, and the design of exception handlers for more robust programs.

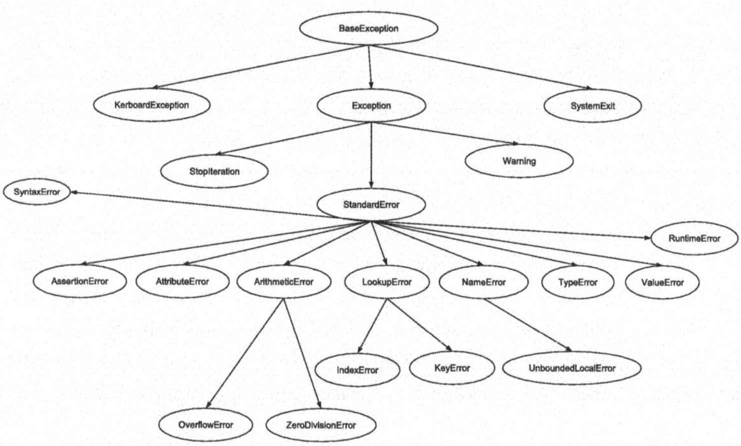

Fig. 10.5 The type hierarchy of Python exceptions

10.3.1 Exception Handling

When exceptions occur, the control flow of the current program is interrupted, and a special control flow sequence is executed. Just to recap, so far in this book, four types of control flow has been introduced, the first three being the three basic types of control flow, including sequential, branching and looping structures, and the fourth being function calls, which can be nested in a stack structure. The control flow of exception handling is different from the four, and is the last type of control flow for Python, and most other imperative programming languages.

When an exception occurs, Python stops the current execution sequence and starts an *exception control flow*. If the current execution sequence is in the global scope, Python stops the whole program, and reports the error. This is the case for the exceptions seen in this book, such as those in Chap. 2. If the current execution sequences not in the global scope, it must be in the local scope of a function. In this case, Python returns to the caller immediately. The same can happen to a stack of nested function calls, leading to a sequence of returning operations, until the global scope is reached, and the program terminates. For example,

```
>>> def f():
...     z=0
...     z+=g(0)
...     return z
...
>>> def g(x):
...     return 1.0/x
...
>>> f()
Traceback (most recent call last):
  File "<stdin>", line 1, in <module>
  File "<stdin>", line 3, in f
  File "<stdin>", line 2, in g
ZeroDivisionError: float division by zero
```

In the example above, IDLE calls *f()*, which in turn calls *g()*, which causes a *division-by-zero* exception. When this happens, the calling stack contains IDLE→*f*→*g*, and the current scope is *g*'s local scope. Python enters exception control flow, and unrolls the stack to *f* and then to IDLE, the global scope. At this point the program is stopped, and the error message is shown, with a traceback of the unrolled calling stack.

The function stack unrolling process can be avoided by **guarding** a block of code against exceptions, and defining an **exception handler**. Python allows exception handling through the **try statement**, the syntax of which is shown in Fig. 10.6. A *try* statement starts with the keyword *try*, followed by a colon(:), and then an indented statement block. The statement block under the *try* line is **guarded** against certain exceptions, indicated by the *except* line, which begins with the keyword *except*, followed by an optional exception specification, and then a colon. The exception specification can be omitted, in which case all exceptions are handled; it can also be a single type name or a tuple of multiple type names, which specify the types of

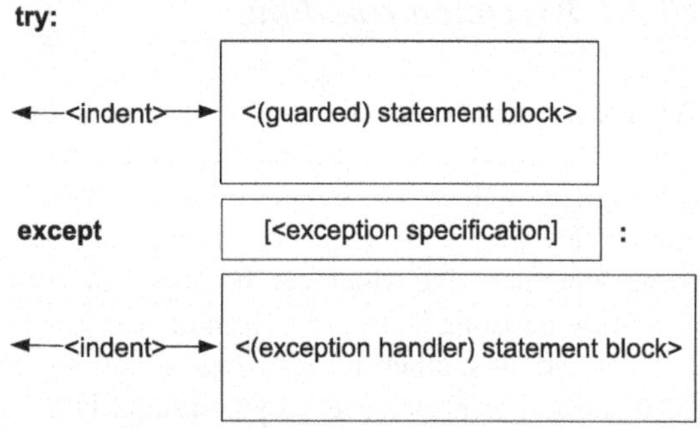

Fig. 10.6 The *try* statement

exceptions to be handled. The exception handler is given in the indented statement block under the *except* line. In the dynamic execution of this statement, the guarded statement block is executed. If no exception happens, the exception handler is never executed. Otherwise, the current execution sequence is stopped immediately, and exception types in the *except* line are checked. If the current exception matches an exception type in the specification, the exception handler is executed, and the exception control flow stops, without the calling stack being unrolled.

To prevent the program from exiting on a zero division error, the function *g* can guard the division expression against the error, returning 0 if it happens.

```
>>> def f():
...     z=0
...     z+=g(0)
...     return z
...
>>> def g(x):
...     try:
...         return 1.0/x # guarded
...     except ZeroDivisionError:
...         return 0 # exception handler
...
>>> f()
0
```

In the example above, expression *1.0/x* in *g* is guarded by a *try* statement, and the zero division error is by the caught *except* line. Execution of the exception handler leads to leads to *g* returning 0, but *f* and IDLE are not affected.

If the exception handler is placed in *f* rather than *g*, the dynamic execution can be different.

```
>>> def f():
...     z=0
...     try:
...         z+=g(0)
...     except:
...         pass
...     return z
...
>>> def g(x):
```

```
...     return 1.0/x
...
>>> f()
0
```

The code above yields the same results as the previous example. However, the dynamic execution sequence is different. When the zero division error occurs, *g*'s execution is interrupted, and the exception control flow starts, leading to an immediate return of *g*, unrolling the function call stack to *f*. However, the line *z+=g(0)*, at which the exception happens in *f*, is guarded against all exceptions. As a result, the exception handler is executed, and the control flow goes back from exception mode to normal.

It is possible to define different exception handlers for different exceptions, by writing multiple *except* components in a compound *try...except* statement. For example

```
>>> def f(x):
...     try:
...         print 1.0/x
...     except ZeroDivisionError:
...         print 'Division by zero'
...     except (TypeError, ValueError):
...         print 'Type or value error'
...
>>> f(0)
Division by zero
>>> f('abc')
Type or value error
```

In the example above, the statement *print 1.0/x* is guarded against three types of exceptions. Two exception handlers are defined under two different *except* lines, the first catching zero division errors, while the second catching type and value errors. Two different components can be added to a *try* statement, including an *else* line, which leads an indented statement block that is executed if no exception happens, and a *finally* line, which leads an indented statement block that is executed regardless whether exceptions happen or not. For example

```
>>> def f(x):
...     try:
...         print 1.0/x
...     except ZeroDivisionError:
...         print 'Division by zero'
...     else:
...         print 'okay'
...     finally:
...         print 'finishing up'
...
>>> f(0.0)
Division by zero
finishing up
>>> f(1.0)
1.0
okay
finishing up
```

The *finally* component is useful for releasing external resources that are used by program, such as files or network connections, no matter whether exceptions occur.

10.3.2 Exception Objects

It is possible to catch the exception instance in an *except* clause, by using the exception specification *<exception type> as <identifier>*:

```
>>> try:
...     f=open('does not exit.txt')
... except IOError as e:
...     print 'IO error, no %d, %s'%(e.errno, e.strerror)
...
IO error, no 2, No such file or directory
```

The example above attempts to open a non-existent file in a guarded statement block, which causes an IO error. The exception is caught, and assigned to the identifier *e*. An IO error instance has two attributes, *errno* and *strerror*, which store the error number and messages, respectively. In general, exception instances are very simple, baring necessary information about the error. In most cases, a string conversion of an exception instance shows all the information.

Custom exceptions can be defined by extending an exception type in the type hierarchy. For example, suppose that a custom function requires an integer type input, and raises an error when the input type is incorrect. One can define a specific type error by extending the class *TypeError*.

```
>>> class MyTypeError(TypeError):
...     def __init__(self, val):
...         self.val=val
...     def __str__(self):
...         return 'Integer expected, but %s received'%self
    .val
...
```

The new class, *MyTypeError*, supports a constructor and a *__str__* method. A *MyTypeError* instance uses an attribute *val* to keep the incorrect input, and shows it in a string conversion.

An exception instance can be raised by using the **raise** statement, which triggers the exception control flow. For example,

```
(continued from above)
>>> def f(x):
...     if type(x)!=int:
...         raise MyTypeError(x)
...     return 1.0/x
...
>>> f('abc')
Traceback (most recent call last):
  File "<stdin>", line 1, in <module>
  File "<stdin>", line 3, in f
```

```
__main__.MyTypeError: Integer expected, but abc
   received
```

MyTypeError can also be guarded against in a *try* statement:

```
(continued from above)
>>> def g(x):
...     try:
...        print f(x)
...     except ZeroDivisionError as e:
...        print 'Zero division error:', e
...     except TypeError as e:
...        print e
...
>>> g(0)
Zero division error: float division by zero
>>> g(1.0)
Integer expected, but 1.0 received
```

In the example above, the expression $f(x)$ is guarded against zero division errors and type errors. Because *MyTypeError* is a sub type of *TypeError*, it is caught by the clause *except TypeError as e*. The string conversion of *e* is polymorphic in that the sub class method *MyTypeError.__str__* is executed dynamically, although the static code *except TypeError as e* only tells that *e* is a *TypeError*, which is the base class.

Exercises

1. Some matrix operations in Chap. 9 has not been included in the *Matrix* class in this chapter. Make the class more powerful by saving it into *Matrix.py*, letting it inherit the *object* class, and
   (a) allow matrix to matrix multiplication (e.g. *m1 * m2*);
   (b) allow in-place addition (e.g. *m1* + = *m2*), in-place scaling (e.g. *m1 *= 3*) and in-place multiplication (e.g. *m1 *= m2*);
   (c) allow matrix transposition (e.g. *m.transpose()*).
2. Write a class hierarchy for simple polygons.
   (a) Write a class, *Triangle*, which is constructed using six argument, x_1, y_1, x_2, y_2, x_3 and y_3, specifying its three vertexes. The class has two methods, *__str__* and *area*. The *__str__* returns a string in the format of 'Polygon: (x_1, y_1), (x_2, y_2), (x_3, y_3)', and the *area* method returns the area of the triangle. Both methods take no arguments.
   (b) Write a class, *Quadrilateral*, which is a sublass of *Triangle*. Try to use as few additional attributes as possible. The constructor of *Quadrilateral* takes eight arguments, x_1, y_1, x_2, y_2, x_3, y_3, x_4 and y_4, specifying its four vertexes. The class has its own *__str__* and *area* methods, with similar meanings as those of *Triangle*.
   (c) Write a class, *Pentagon*, by sub classing *Quadrilateral*.
   (d) Write a class, *Hexagon*, by sub classing *Pentagon*.

(e) Add a method, *regular*, to the polygon class, which takes no argument and returns whether the polygon is regular.
(f) [Challenge] Add a method, *convex*, to the polygon classes, which takes no argument and returns whether the polygon is convex.

3. (a) Write a class, *Student*, which has three properties, including a 6-digit student ID, a name, and a list of exam scores. The constructor takes the student name as the first argument, followed by an optional number of additional arguments representing the examination scores. The *Student* class provides two methods, including *average*, which takes no argument and returns the average score of the student, and *__str__*, which formats the student information as '[ID\t name\t scores]', where *scores* is a tab-seperated list of floating-point numbers, accurate to two digits after the decimal point. For example,

```
>>> s=Student('John Smith')
>>> print s
100001   John Smith
>>> s=Student('Mary Harper', 100, 90.5, 75)
>>> print s
100002 Mary Harper 100.00 90.50 75.00
```

Note that the student IDs are allocated automatically.
(b) Write a new class, *Class*, which has a single attribute, recording a list of students. The class supports three methods, *load(path)*,which loads the list of students from a text file at *path*, *add(student)*, which adds a new student instance to the list, and *save(path)*, which saves the list of student into a text file. The format of files is one student per row, in the string conversion form.
(c) The questions (a) and (b) define a rudimentary student database. Try to make it more useful by adding funtionalities. For example, add a new method, *add_score*, to the *Student* class, which takes a single argument representing a score, and adds it to the list of scores of the student.

4. Extend the functionalities of the *Account* class by adding a new class, *AccountIO*, which acts as a text interface to accounts.

(a) Make the *AccountIO* class, which contains an attribute, *accounts*, maintaining a list of *Account* instances. The constructor initializes *accounts* to [ ].
(b) Add a *formstr* method to *Account*. It takes a string input, in the format of *Account.__str__* output, assigning to the *Account* instance the corresponding attributes.
(c) Add serialization to *AccountIO*. Modify the constructor so that it takes a string argument specifying a file path. The file can be either empty, or contain the information of a *list* of accounts. The constructor initializes the attribute *accounts* according to the file, and saves the path as an attribute. Add another method, *save*, which saves the *list* of accounts into the file path specified in the constructor call.

(d) Add a new method, *add*, to *AccountIO*, which takes a single account instance and adds it to *accounts*.
(e) Add a new method, *find*, to *AccountIO*, which takes two arguments specifying the account holder name and the account number, respectively, returning the corresponding instance from *accounts*, or *None* if there is not such a record.
(f) Add a new method, *run*, to *AccountIO*. It repeatedly asks the user for commands via text IO. Allowed commands include 'new account', the system asks for the user name and initial balance. For 'login', the system asks for user name and account number. If login is successful, the system can display balance, perform deposit and withdrawl and perform logout.
(g) Modify the *Account* class, raising exceptions when invalid deposit or withdraw is performed. catch the exceptions in *AccountIO*.
(h) [Challenge] The class *AccountIO* above does not consider the consistency of account numbers. This is because there is no guarantee that a new account does not take the number of accounts in the record file. Modify the *Account* and/or *AccountIO* classes so that no two accounts have the same number.

5. Draw out the memory structures when the following statements are executed based on the setting of Fig. 10.3, explaining the change in each step of the process.

   (a) When *a2 = CreditAccount('Merina',limit=500)* is executed;
   (b) When *a3 = Account('Catherine', 1000)* is executed;
   (c) When *a2.deposit(100)* is executed;
   (d) When *a2.withdraw(500)* is executed.

6. State the three most important elements of object-oriented programming.

Chapter 11
Summary

11.1 The Structure of a Python Program

A **Python program** is a source text file with the extension name '.py'. It consists of a sequence of *statements* (Chap. 2), which are executed one after another when the Python program is executed. A statement can be *compound* (Chap. 4), which contains statement blocks in itself. The *dynamic execution sequence* of a compound statement can be different from its static form, involving branches (Chap. 4), loops (Chap. 4), function calls (Chap. 6) and exception control flow (Chap. 10).

The structure of a Python program is summarized in Fig. 3 of Chap. 3. The basic building block of Python statements are *key words* (Chap. 2), identifiers (Chap. 2), literals (Chap. 2) and operators (Chap. 2). They can be combined to form *expressions*, which are *value*-carrying elements in statements.

11.1.1 Expressions

Expressions represent values, which are ultimately objects in Python. The simplest expressions can consist of a single literal (e.g. 123 and *None*) or identifier (e.g. *x*, *_abc* and *y0*). More complex expressions can consist of recursive combinations of sub expressions using operators (e.g. +, *, [ ] and call). In the execution of a program, python *evaluates* every expressions to reduce it to an object in memory, which belongs to a *type* , and has a *value*. For example, when the expression *3.0* is evaluated, a *float* object 3.0 is constructed to represent its value. When the expression (3 + 5) is represented, the literals 3 and 5 are first realized into *int* objects in memory, and then the '+' operator is applied, resulting in the object 8, which represents the value of the expression.

Chapter 3 gives a summary of types, literals, operators and statements up to that point in the book. Many more types, operators and statements have been introduced after Chap. 3. Table 11.1 gives a comprehensive summary of Python

Y. Zhang, *An Introduction to Python and Computer Programming*,
Lecture Notes in Electrical Engineering 353, DOI 10.1007/978-981-287-609-6_11

Table 11.1 Python types and literals

Type	Literal	Common operators	Mutable	Comment
int	$1, 0, -1$	$+, -, *, /, \%, **,$ $<<, >>, \backslash, \hat{}, !$	No	Integer
long	$1L$	same as above	No	Implicit large integer
float	$3.5, 5e-1$	$+, -, *, /, \%, **$	No	Floating point number
complex	$1-3j$	$+, -, *, /, \%,$ methods	No	Complex numbers
str	'abc', "1!"	$+, *,$ methods	No	strings
bool	True, False	and, or, not, $>, >=,$ $<, <=, ==, !=,$ is, is not, in, not in	No	Boolean values
tuple	(), (1,), (1,2.0,'a')	$+, *,$ [], methods	No	Sequence collection object
list	[], [1], [1, 3.0, '*abc*']	$+, *,$ [], methods	Yes	Sequence collection object
dict	{}, {1 : '*a*', 2 : '*b*', 3 : '*c*'}	[], methods	Yes	Mapping collections object
set	set(), set([1, 2, 3])	$\&, \vert, \hat{}$ methods	Yes	Set collection object
None type	None		No	Nothing

types and literals, which include number types (*int*, *float* and *complex*), strings (*str*), Boolean values (*bool*), sequential collection types (*tuple* and *list*), mapping collection types (*dict*) and sets (*set*). In addition, there are several special types of objects, including functions (the *def* statement), which support the call (i.e. the () operator), modules (the *import* statement), which support the dot (i.e.) operator. Finally, custom types can be defined using the *class* statement. Each custom type can be treated as a separate type, more about which will be discussed in the next section.

A list of **operators** in Python is shown in Table 11.2, which is a fully-extended version of Table 2 in Chap. 4. Each operator takes a certain number of *operands*, which have specific types, and results in a value of a specific type. For example, the + operator can take two integers and result in an integer. The > operator can take two integers, and result in a Boolean value. The $*$ operator can take a string and an integer, and result in a string (i.e. string repetition). Regardless whether an expression is simple(e.g. single literal) or complex(e.g. $3+5*int(raw\_input('x='))$), Python treats it as a single object. For example, consider the following case:

```
>>> x ='a'
>>> e = [x, 123, 3*3>8.8]
>>> e
['a', 123, True]
```

The literal of *e* consist of three elements, the first being an identifier, the second being a literal, and the third consists of operators ad literals. In the evaluation of the list literal, all of them are reduced to single objects.

Table 11.2 Python operators

Operator	Comment
+, −, *, /, %, **, //	Arithmetic operators
<<, >>	Bitwise shift operators
&, |, ^, !	Bitwise logic operator
.	Get identifier from module Get attributes and methods from instance Get attributes and methods from classes
[]	Getitem, getslice
()	Function call
==, ! =, >, >=, <, <=	Arithmetic comparisons
and, **or**, **not**	Boolean operators
is, **is not**	Identify comparisons
in, **not in**	Membership comparisons
&, |, ^	Set operator
lambda	Lambda functions
x **if** *y* **else** *z*	Ternary operator
[**for** ... **in** ... **if**...]	List comprehension

Python offers a builtin function, *eval*, which takes a string that specifies an expression, and returns an object that represents the result of its evaluation. For example,

```
>>> eval('1.0') # a literal
1.0
>>> eval('5*5') # operator result
25
>>> eval('"abc"+"def"+"ghi"')
'abcdefghi'
>>> a=1
>>> eval('a') # identifier
1
>>> eval('"x"') # string literal
'x'
>>> eval('b') # undefined
Traceback (most recent call last):
  File "<stdin>", line 1, in <module>
  File "<string>", line 1, in <module>
NameError: name 'b' is not defined
>>> f=eval('lambda x: x+1') # lambda function
>>> eval('f(1)') # function call
2
>>> x = eval(raw_input('x=')) # built-in function call
x=1.1,3.5,3.9
>>> x
(1.1, 3.5, 3.9)
```

```
>>> eval('c=1') # not an expression
Traceback (most recent call last):
  File "<stdin>", line 1, in <module>
  File "<string>", line 1
    c=1
     ^
SyntaxError: invalid syntax
```

In the example above, several expressions are written as strings, and passed as the only argument to the function *eval*, which calculates the values of the expressions and returns the corresponding objects. In fact, the IDLE program also calls *eval* in order to obtain the result for evaluating the expression. In the last line, 'c=1' is a statement rather than an expression, and therefore cannot be evaluated by *eval*.

In the opposite direction, when IDLE shows the value of an expression to the user, it calls the built-in function *repr*, which takes a single object argument and returns a string that contains the **internal representation** of the argument. For example,

```
>>> a=1.0
>>> repr(a)
'1.0'
>>> repr(1.0/7)
'0.14285714285714285'
>>> print 1.0/7 # this is str(1.0/7)
0.142857142857
>>> c='abc\ndef\nghi'
>>> repr(c) # internal representation
"'abc\\ndef\\nghi'"
>>> print c # stv representation
abc
def
ghi
>>> f=lambda x: x+1
>>> repr(f)
'<function <lambda> at 0x1006ef398>'
>>> class A:
...     pass
...
>>> repr(A)
'<class __main__.A at 0x1006c8b48>'
>>> a=A()
>>> repr(a)
'<__main__.A instance at 0x1006e9f38>'
```

Each Python object has its own *internal representation*. Although both being *str* objects, the internal presentation of built-in objects can be different from their *string representation*, with the former being more precise and the latter being more formatted.

Instances of custom types also have a default internal representation. Similar to the string representation, which can be customized by using the *__str__* method, the internal representation of an instance can be customized using the special method *__repr__*.

```
>>> class A:
...     def __repr__(self):
...         return 'I am A.'
...     def __str__(self):
...         return 'A'
...
>>> a=A()
>>> repr(a)
'I am A.'
>>> str(a)
'A'
>>> a # IDLE shows repr(e) given an expression (e)
I am A.
>>> print a # default string representation
A
```

11.1.2 Statements

A **statement** is the basic unit of execution in a Python program. Each statement achieves a particular functionality, and has a fixed underlying mechanism of execution. For example, the assignment statement binds an identifier to a value object, the *print* statement shows the string representation of a value on the text console, and the *def* statement constructs a function object, binding it to an identifier. Statement can be simple (e.g. assignment and *print*) or compound (e.g. *if* and *def*), containing a block of sub statements. Compound statements can also be nested (e.g. nested loops), resulting in complex dynamic execution sequences. A summary of statements in Python is shown in Table 11.3, which is a full-extended version of Table 3 in Chap. 2.

Python provides a built-in function, *exec*, which takes a single string argument, representing a statement, and executes the statement. The function returns *None*. For example,

```
>>> exec('c=1') #assignment
>>> c
1
>>> exec('print "abc"') #print
abc
>>> exec('import math') #import
>>> print math.pi
3.14159265359
>>> exec('while c<7:\n\tc+=2\n\tprint c') #compound,
   tab indentation
3
5
7
>>> exec('print c; c=0') #semi-colon
7
>>> exec('print c\nc+=1') #two lines
0
```

Table 11.3 Summary of statements in Python

Statement	Syntax
assignment (Chaps. 2, 7, 10)	<identifier> = <expression>
	<identifier> += <expression>, ...
del (Chaps. 2, 7, 10)	**del** <identifier>
import (Chap. 3)	**import** <module>
	import <module> **as** <identifier>
	from <module> **import** <identifier>
print (Chap. 3)	**print** <string> [, <string> ...]
if (Chap. 4)	**if** <bool>:
	<statement block>
	elif <bool>:
	<statement block>
	else:
	<statement block>
while (Chap. 4)	**while** <bool>:
	<statement block>
	else:
	<statement block>
for (Chaps. 5, 8, 9)	**for** <identifier> **in** <iterable>:
	<statement block>
	else:
	<statement block>
break (Chap. 4) continue (Chap. 4)	**break**
	continue
def (Chap. 6)	**def** <identifier> (<parameter(s)>):
	<statement block>
pass (Chap. 6)	**pass**
return (Chap. 4)	**return** <expression>
yield (Chap. 9)	**yield** <expression>
class (Chap. 10)	**class** <identifier> [(<parameter(s)>)]:
	<statement block>
try (Chap. 10)	**try**:
	<statement block>
	except <exception specification>:
	<statement block>
	else:
	<statement block>
	finally:
	<statement block>
raise (Chap. 10)	**raise** <Exception>

Given one or multiple lines of user input, IDLE treats it as a string, and calls *exec* to execute it as a simple or compound statement.

11.2 The Data Model of Python

The term **syntax** is commonly used to describe the structure of Python programs. The statement structures summarize in the last section describe the correct syntax of a Python program. A program that violates the correct syntax will lead to a **syntax error** during execution. For example, the *if* statement requires a Boolean expression (e.g. *c==1*) in the *if* line; if a statement (e.g. *c=1*) is given in the place, a syntactic error will be raised in the execution. Knowing the correct syntax is the first step in writing valid programs.

On the other hand, the term **semantics** refers to the meaning of statements, or the underlying mechanism of execution. Understanding the semantics of Python statements is as important as understanding its syntax for programming. The most important aspect of semantics is the **data model**, or how information is represented and processed by Python statements. As summarized in the last section, Python uses **objects** to represent all types of information. Objects are central to Python's data model, and hence its semantics.

11.2.1 Identity, Type and Value

Each Python object has an **identity**, a **type** and a **value**. The *identity* of an object is its memory address, which can be obtained using the *id* function introduced in Chap. 2. The identity of an object is fixed once the object is constructed in memory, and never changes throughout the object's existence. The *type* of an object can be obtained using the *type* function; it determines its range of values, operators and attributes. Common types can be classified into numbers (e.g. *int*, *long*, *float*, *complex*), sequence (e.g. *str*, *type*, *list*), mappings (e.g. *dict*), sets (e.g. *set*), callables (e.g. functions, methods), models, classes, instances and files. In addition to these types, information that is used internally by Python is also represented by specific types of objects in memory. For example, Python code is represented by the *code* type; the statements and list of arguments in function objects (Chap. 6) are coded in code object. Slices (Chap. 3) and Tracebacks (Chap. 10) are also represented by internal types. Similar to the identity, type of an object cannot be changed once the object is constructed in the memory.

Among the identity, the type and the value, the only thing that can possibly be changed for an object is its value, and only objects of mutable types can change their values. The value of a mutable object can be changed using specific operators (e.g. *setslice*) or methods (e.g. *append*), but not using the assignment statement (e.g. *l=[1,2,3]*). The value of an immutable object cannot be changed once it is constructed.

Mutability can sometimes be confused with **assigment**, which changes **binding** between identifiers and objects. Identifiers are names to objects in Python; they are maintained in binding tables. Python looks up identifiers in a scope hierarchy, which includes the local (function) scope, the global (module) scope and the built-in scope. Details of binding, scopes and mutability can be found in Chaps. 3, 6 and 7, respectively.

11.2.2 Attributes and Methods

In addition to an identity, a type and a value, an object can also have a set of **attributes** and **methods**, which can be obtained using the *dir* function introduced in Chap. 3. For example, complex numbers have the attributes *real* and *imag*, and custom type instance can have arbitrary attributes and methods. Methods can be regarded as a special type of callable attributes.

Most built-in types do not allow **user-defined** attributes, which instances of custom types do. For example, Chap. 10 shows how arbitrary attributes, such as *balance*, can be defined for an *Account* instance. In contrast, complex numbers do not allow such attribute definitions. The attributes of complex numbers are **built-in**, or hard-coded, as a par of the special method *__getattr__*. Attributes of most built-in types are read-only, and cannot be modified using the *setattr* method. Although not frequently used, functions also allow user-defined attributes:

```
>>> f=lambda x,y:x+y
>>> f.a=1
>>> f.a
1
```

The implementational reason for instances and functions to support user-defined attributes is that they are associated with their own binding tables. The binding table of an object is implemented as a special attribute of the object named *__dict__*. The content of the binding table can be accessed from the *__dict__* attribute which stores identifiers (i.e. attributes of the object) as keys and their values as values. For example, the user-defined attributes of the function *f* above can be accessed by using

```
>>> f.__dict__
{'a': 1}
```

As introduced in Chap. 3, models also have binding tables associated with them. The binding table of a module object is also implemented as a special attribute named *__dict__*, which contains all the identifiers in the module and their values as items. Because of the *__dict__* attribute, it is also possible to define arbitrary identifiers in a module. For example,

```
>>> import math
>>> math.a=1
>>> math.a
1
>>> math.__dict__ # or dir(math)
```

```
{'pow': <built-in function pow>, 'fsum': <built-in
    function fsum>, 'cosh': <built-in function cosh>, '
    ldexp': <built-in function ldexp>, 'hypot': <built-
    in function hypot>, 'acosh': <built-in function
    acosh>, 'tan': <built-in function tan>, 'asin': <
    built-in function asin>, 'isnan': <built-in
    function isnan>, 'log': <built-in function log>, '
    fabs': <built-in function fabs>, 'floor': <built-in
     function floor>, 'atanh': <built-in function atanh
    >, 'modf': <built-in function modf>, 'sqrt': <built
    -in function sqrt>, '__package__': None, 'frexp': <
    built-in function frexp>, 'degrees': <built-in
    function degrees>, 'lgamma': <built-in function
    lgamma>, 'log10': <built-in function log10>, '
    __doc__': 'This module is always available.  It
    provides access to the\nmathematical functions
    defined by the C standard.', 'asinh': <built-in
    function asinh>, 'fmod': <built-in function fmod>,
    'atan': <built-in function atan>, 'factorial': <
    built-in function factorial>, '__file__': '/Library
    /Frameworks/Python.framework/Versions/2.7/lib/
    python2.7/lib-dynload/math.so', 'copysign': <built-
    in function copysign>, 'expm1': <built-in function
    expm1>, 'ceil': <built-in function ceil>, 'isinf':
    <built-in function isinf>, 'sinh': <built-in
    function sinh>, '__name__': 'math', 'trunc': <built
    -in function trunc>, 'a': 1, 'cos': <built-in
    function cos>, 'pi': 3.141592653589793, 'e':
    2.718281828459045, 'tanh': <built-in function tanh
    >, 'radians': <built-in function radians>, 'sin': <
    built-in function sin>, 'atan2': <built-in function
     atan2>, 'erf': <built-in function erf>, 'erfc': <
    built-in function erfc>, 'exp': <built-in function
    exp>, 'acos': <built-in function acos>, 'log1p': <
    built-in function log1p>, 'gamma': <built-in
    function gamma>}
```

In summary, attributes and methods are implemented in two ways. Custom attributes and methods are stored in the *__dict__* attribute of an object, and can be defined, changed or deleted arbitrarily. Built-in attributes and methods can be hardcoded, such as using the special *__getattr__* method; and are read-only. The built-in function call *dir(x)* also shows all the keys of *x.__dict__* if the attribute exist, and tries to list out all the built-in attributes.

11.2.3 Documenting Objects

Comments are useful in ensuring the readability of Python code. Chapter 3 introduces the use of **in-line** and **end-of-line** comments, which can be regarded as two ways to add documentation to abstract programs. In addition to comments, Python supports another type of documentation.

Docstrings. Docstrings are documentation strings for modules, functions, classes and methods. They must be written as a string constant expression, and placed as the first statement in the definition of an object. For modules, the docstring should be written as the first line in a Python file. For classes, the docstring should be written as the first line of the class body. The same applies to functions and methods.

Different from comments, which are ignored at runtime, docstrings are attached as a special *__doc__* attribute to their objects, ence being able to support online documentation and interactive help systems. For example,

```
[doc.py]
"""This is a module. A second line to explain the
   usage. One third line of documentation."""
class A:
    "This is a class"
    def f(self):
        "This is a method"
def f():
        "This is a function"
```

The module *doc.py* consists of a class and a function, with a method defined in the class. The module, class, function and method are all associated with docstrings.

```
>>> import doc
>>> doc.__doc__
' This is a module. A second line to explain the usage
   . One third line of documentation.'

>>> doc.A.__doc__
'This is a class'

>>> doc.A.f.__doc__
'This is a method'

>>> doc.f.__doc__
'This is a function'
```

Th documentation string of the objects can also be accessed using the built-in function *help* interactively.

```
>>> import doc
>>> help (doc)
NAME
    doc - This is a module. A second line to explain
        the usage. One third line of documentation.

FILE
    /Users/yue_zhang/doc.py

CLASSES
    A

    class A
     |   This is a class
     |
     |   Methods defined here:
```

```
        |
        |   f(self)
        |       This is a method

FUNCTIONS
    f()
(END)

>>> help (doc.A)
Help on class A in module doc:

class A
 |  This is a class
 |
 |  Methods defined here:
 |
 |  f(self)
 |      This is a method
(END)

>>> help (doc.A.f)
Help on method f in module doc:

f(self) unbound doc.A method
    This is a method
(END)

>>> help (doc.f)
Help on function f in module doc:

f()
    This is a function
(END)
```

11.3 Modules and Libraries

In practice, a software application typically consists of multiple source files. A Python project typically consists of multiple modules. **Modules** correspond to Python files, except for **built-in** modules, which are written as a part of the Python application program. As shown in Chap. 3, there two ways to execute a Python file, one being execution as the main program, and the other being execution by importation. At runtime, the two ways of execution result in a **main module** and a **library module**, respectively.

At runtime, each module object has an attribute *__name__*, which uniquely identifies the module object. *__name__* is an identifier in the binging table of all modules. The example below illustrates example values of *__name__*:

```
>>> import math
>>> math.__name__
'math'
```

```
>>> import random as m
>>> m.__name__
'random'
```

When executed as the main program, the value of *__name__* of a module is set to '*__main__*'. This special value can be expoited for conditionally executing ome code only when the module is executed as the main program. For example, the file *mul.py* below provides a function, *prod*, which takes a list arguement and returns the product of all the items in the list.

```
[mul.py]
def prod(e):
    retual = 1
    for i in e:
        retval *= i
    return retval
if __name__ == '__main__': # when executed as the main
    program
    s = raw_input('Enter a list of numbers')
    s = '[' + s +']'
    print 'The product is: ', prod(eval(s))
```

When imported as a module, *mul.py* does not perform any operations except for defining the function.

```
>>> import mul
>>> mul.prod([1,3,6])
18
```

However, when executed as the main program, *mul.py* asks th user to enter a sequence of numbers, and then returns their product.

```
Zhang's Macbook Pro ~ Desktop$ python mul.py
Enter a list of numbers: 1, 3, 5, 7, 9
The product is: 945
```

The condition *__name__* == '*__main__*' is commonly used when some functionalitie must be conditioned on the running mode.

11.3.1 Packages

In large projects, multiple module can be written for performing the same functionality. The module files are typically put into the same folder. Conceptually, a set of modules under the same big task forms a **package**. In Python, a **package** is a special module that can contain sub modules. For example, there is a module *os* which is a package that contains methods that communicate with the operating system. Under *os* there is a sub module *os.path*, which contains methods that are related to the folder hierarchy. the function *os.path.abspath* is a function that takes a string argument that indicates a relative path, and returns a string that indicates the corresponding absolute path.

A package can be implemented as a folder that contains a Python file with the special name *__init__.py*. When the package is imported, *__init__.py* is excuted as the module file that is imported, with every identifier being saved into the binding table of the package. Sub modules can be put into the folder, and automatically imported into its binding table via *__init__.py*. For example, the folder *pac* below is a package, having a file *__init__.py*.

```
pac /
      __init__.py
      mod1.py
      mod2.py
```

The content of *__init__.py* can be:

```
[__init__.py]
a = 1
def f(x):
      return x+1
import mod1
import mod2
```

The content of *mod1.py* can be:

```
[mod1.py]
b = 2
def g(x):
      return x+2
```

The program below illustrates usage of the package:

```
>>> import pac
>>> pac.f(1)
2
>>> pac.mod1.g(pac.a)
3
```

Packages can also form hierarchies, with sub packages containing sub modules. Thay are not commonly necessary, and interested readers can refer to Python documentation for more details.

11.3.2 Library Modules

A number of standard library modules have been inttroduced in this book, including the *math*, *cmath* and *random* modules introduced in Chap. 2, the *__builtin__* module introduced in Chap. 3, the *heapq* module introduced in Chap. 6, the *copy* module introduced in Chap. 7, and the *urllib* module introduced in Chap. 9. Library modules greatly facilitate Python programs in achieving complex functionalities, ranging from scientific computing to game design, form local database management to web service, and from text I/O to multimedia user interface. Their availability helps Python in becoming one of the most popular programming languages.

A detailed introduction of common Python libraty modules is beyond the scope of this book, the goal of which is to introduce the Python language and basic programming thinking to readers that are new to computer science. Library modules typically require background knowledge of specific area in computer science, such as algorithms, databases, networks, computer audio, multimedia, operating systems and computer security, which are difficult to cover in short chapters.

Table 11.4 summaries commonly used Python modules and external libraries for interested readers. The modules are organized according to their functionalites, with very brief introductions given for each mudule. More information can be obtained from the standard Python documentation. External modules are marked out, but no URLs are given. The latest website can be accessed via web search using the name of the library modules. Finally, because the developer community is highly active, the most popular library modules can frequently change, and it is therefore useful to constantly look for the latest tools.

Table 11.4 Python modules and external libraries

Functionality	Module	Description	External
Python system	__builtin__	the built-in scope	
	argparse	Command-line option and argument parsing	
	atexit	Register and execute cleanup functions	
	bdb	Debugging framework	
	disvtils	Support for building, Python installation packages	
	exceptions	Standard exception classes	
	gc	Interface to the garbage collector	
	getopt	Parser for command-line options	
	hotshot	A profiler to measure runtime performance	
	imp	Access the implementation of the import statement	
	importlib	Customize import functionalities	
	inspect	Extract information and source code from live objects	
	io	Core stream tools	
	keyword	Test whether a string is a keyword	
	modulefinder	Find modules used by a script	
	operator	Functions corresponding to standard operators	
	os	Operating system interfaces	
	pdb	The Python debugger for interactive interpreters	
	pkgutil	Utilities for the import system	
	pprint	Functions for pretty print	
	profile	Python source profiler	
	pstats	Statistics object for use with the profiler	
	pydoc	Documentation and online help	

(continued)

Table 11.4 (continued)

Functionality	Module	Description	External
	runpy	Locate and run Python modules without importing them	
	sys	Access runtime system parameters and functions	
	sysconfig	Python's configuration information	
	test	Regression tests package	
	thread	Create multiple threads	
	threading	High-level thread interface	
	timeit	Measure the execution time of small code	
	trace	Teace code execution	
	traceback	Print function stack	
	unittest	Unit testing framework	
	warnings	Waring messages	
	zipimport	Importing modules from ZIP archives	
	nose	A testing framework	Yes
	IPython	An enhancement of IDLE	Yes
Data types and algorithms	array	space efficient arrays	
	calendar	Calendars	
	cmath	Complex numbers	
	collection	Hign-performance data typies	
	copy	Shallow and deep copy functions	
	datatime	Date and time	
	diftlib	Functions to compute different between objects	
	encodings	String encoding/Unicode	
	fractions	Rational numbers	
	functools	Higher-order functions	
	gettext	Multilingual internationalization service	
	headq	Priority queue	
	itertools	Iterators for looping	
	locale	Internationalization service	
	marshal	Convert Python objects to byte streams	
	math	Mathematical functions	
	numbers	Numerical abstract base classes	
	random	Random numbers	
	re	Regular expressions	
	string	Common string operations	
	StringIO	Read and write string as files	
	struct	Interpret strings as packed binary data	
	textwrap	Text wrapping and filling	
	time	Time acccess and conversions	

(continued)

Table 11.4 (continued)

Functionality	Module	Description	External
	types	Built-in types	
	userDict	Class wrapper for dict objects	
	userList	Class wrapper for list objects	
	weakref	Weak references and dictionaries	
	numpy	Math functions	Yes
	scipy	Scientific computation	Yes
	matplotlib	A numeric plotting library	Yes
	nltk	Languages	Yes
	sympy	Algebra	Yes
Graphical User interface	Tkinter	Interface to Tcl/Tk GUI	
	Tix	Tk extension widgets	
	ttk	Tk widget set	
	Scrolled Text	Tk Text widget with a verticle scroll bar	
	turtle	Tk Turtle graphics	
	wxPython	A GUI system based on wxWidget	Yes
	PyQT	A GUI system based on QT	Yes
	PIL	Python image library	Yes
	Pygame	2D game tools	Yes
	Pyglet	3D game and animation	Yes
Files and databases	bsddb	Interface to Berkeley DB	
	bz2	Bzip2 compression	
	dbhash	DBM-style interface to the BSD database	
	filecmp	Compare files efficiently	
	fileinput	loop over standard input or a list of files	
	gzip	Gzip file compression	
	mmap	Memory mapped files	
	pickle	Convert Python objects to streams	
	pickletools	Configure pickle functions	
	shutil	High-level file operations	
	sqlite3	Interface to SQLite3.X	
	tarfile	Tar archive files	
	tempfile	Temporary files and directories	
	zlib	Compression and decompression	
Web client server	BaseHTTPserver	Basic HTTP server	
	cgi	Common Gateway Interface	
	CGIHTTPServer	CGI HTTP Server	
	cgith	CGI traceback	
	email	Manipulating email messages	

(continued)

Table 11.4 (continued)

Functionality	Module	Description	External
	ftplib	FTP protocol client	
	json	JSON format	
	mailbox	Mailbox	
	poplib	POP3 protocol client	
	SimpleHTTPServer	Simple HTTP Server	
	SimpleXMLRPCServer	Simple XML RPC Server	
	smtpd	SMTP server	
	smtplib	SMTP client	
	socketServer	Socket based servers	
	ssl	SSL socket objects	
	telnetlib	Telnet client	
	urllib	Open arbitrary URL	
	urllib2	Urllib version 2	
	urlparse	Parse URL strings	
	xml	Parse XML	
	xmlrpclib	XML RPC client	
	requests	HTTP library	Yes
	scrapy	Web scraping	Yes
	scapy	A packed sniffer	Yes
	django	Web server framework	Yes

The manufacturer's authorised representative in the EU is Springer
Nature Customer Service Centre GmbH, Europaplatz 3, 69115 Heidelberg,
Germany. If you have any concerns regarding our products, please
contact ProductSafety@springernature.com

Printed and bound by CPI Group (UK) Ltd, Croydon, CR0 4YY
12/12/2025
02018127-0003